大学英语特色课程系列教材

College English for Textile and Fashion Majors (Book Two)

纺织服装专业
大学英语综合教程（第2册）

总主编　胡伟华

主　编　郑卉蓉　　副主编　石发德　李　伟　廖　燕　魏　庆

西安交通大学出版社
XI'AN JIAOTONG UNIVERSITY PRESS

国家一级出版社
全国百佳图书出版单位

图书在版编目（CIP）数据

纺织服装专业大学英语综合教程：College English for Textile and Fashion Majors. 第2册 / 郑卉蓉主编；胡伟华总主编. — 西安：西安交通大学出版社，2022.6
ISBN 978-7-5693-2498-3

Ⅰ.①纺… Ⅱ.①郑… ②胡… Ⅲ.①纺织工业－英语－高等学校－教材②服装工业－英语－高等学校－教材 Ⅳ.①TS1②TS941

中国版本图书馆CIP数据核字(2021)第281482号

纺织服装专业大学英语综合教程（第2册）
College English for Textile and Fashion Majors (Book Two)

总 主 编	胡伟华
主　　编	郑卉蓉
策划编辑	蔡乐芊
责任编辑	庞钧颖　蔡乐芊
责任校对	张　娟
封面设计	任加盟

出版发行　西安交通大学出版社
　　　　　（西安市兴庆南路1号 邮政编码710048）
网　　址　http://www.xjtupress.com
电　　话　（029）82668357 82667874（市场营销中心)
　　　　　（029）82668315（总编办）
传　　真　（029）82668280
印　　刷　陕西思维印务有限公司

开　　本　850mm×1168mm　1/16　印张　19.75　字数　431千字
版次印次　2022年6月第1版　2022年6月第1次印刷
书　　号　ISBN 978-7-5693-2498-3
定　　价　58.90元

前　言
Preface

一、编写背景

本教材以习近平新时代中国特色社会主义思想为指导，是深入贯彻落实全国教育大会和新时代全国高等学校本科教育工作会议精神的成果。本书编写团队根据《教育部关于加快建设高水平本科教育全面提高人才培养能力的意见》，以提升高校专业人才培养质量，践行大学英语教学改革为目标，在"外语+"模式背景下，开发了这套大学英语特色课程系列教材，旨在通过深化英语课程基础性和应用性的结合，服务国家纺织服装专业的人才培养工作，增强国家纺织服装产业在国际社会的话语权。

我们需要培养大学生在各自专业及行业领域里汲取和交流专业信息的能力，以及用英语直接开展工作的能力，因此，亟需能够满足此类需求的相关课程和教材。掌握专业英语对于学生了解快速变化的纺织服装行业尤为重要。

基于此，本书编写团队本着严谨、务实的态度，深入调研相关用人单位和学生的需求，精心策划、编写了这套适合高校纺织服装专业学生使用的大学英语特色教材。

二、编写思路

本教材以《大学英语教学指南（2020版）》为指导，力求准确把握大学英语教学的性质、目标及要求，以及新时代国家和社会对大学生英语水平及专业能力的需求，在设计和编写中体现人文性和工具性，力求全面提升学生的语言水平和综合素养。同时，本教材借鉴近年来大学英语教学改革的成功经验与教学实践的成果，希望通过新的教学思路及设计进一步推动大学英语教学迈向新台阶。

总体来说，本教材在设计和编写中遵循以下原则：

1. 教学目标体现个性化教学的需求。

2. 教学理念体现"教师主导，学生主体"。

3. 教学内容反映时代特色，聚焦热点话题。

4. 教学手段体现数字化和立体化。

三、教材特色

本教材参考借鉴国内外优秀教材，基于混合式教学的实践和学生学情分析编写，具有以下特色。

1. 选材新颖，体现时代和专业特色

本教材主要面向纺织服装相关专业的本科学生，帮助学生在语言习得的同时获取本专业最前沿的知识和信息。鉴于此，本教材所选文章均具有一定的专业特色，同时紧跟时代潮流，力求最大限度地拓展学生的国际视野。这是本教材编写的一个立足点和出发点，也是区别于其他一般教材的特色所在。

2. 思政统领，落实立德树人的根本任务

本教材内容体现思政主题，力求培养学生的家国情怀。具体来讲，本书所设 8 个单元中，每个单元均有反映"中华文化走出去"的内容，例如中国的传统丝绸、中国的品牌、中国纺织行业的发展等，引导学生树立文化自信。

3. 紧密结合大学生学习与生活，趣味性强

美国语言学家克拉申教授在 20 世纪 80 年代初期提出"语言输入假说"(Input Hypothesis)，其核心内容之一就是学生二语习得的材料要遵循"既有趣又相关"的原则，即输入的语言材料越有趣，关联性越强，语言习得的效果越好。因此，本书编写团队在编写前对相关专业学生进行了深入的调研，最终结合教材的内在逻辑和完整性拟定了每个单元的主题。教材所选内容非常贴合大学生的兴趣点，对学生自主学习具有一定的促进意义。

4. 训练学生思辨能力

"新文科"要求打破学科专业壁垒，运用多样化的教育组织方式，抓好教学"新基建"。为更好地培养学生的思辨能力，本教材结合教学实践，在每单元都设置了相应的思辨类话题，题目类型包括思维导图、问卷调查、课堂演讲、课堂辩论等。这些话题紧密结合本单元主题，以多样化的形式丰富了本单元内容，帮助学生拓宽思路，增强思辨能力。

5. 混合式教学模式

混合式教学模式是在多种学习理论的指导下，根据教学内容、学生和教师自身条件，混合传统的面授和网络传授两种课堂形式，以达到预期的学习目标的一种教学模式。本教材提供了丰富的线上线下资料，为混合式教学提供了充分条件，具有很强的实操性。

四、使用建议

本教材旨在有效巩固和扎实培养学生的语言知识技能和专业素养，通过视角多元、内涵丰富、与时俱进的选材，以及形式多样的练习，夯实学生的专业词汇基础，提升学生的专业英语阅读能力，培养学生的国际交流能力。根据专门用途英语阶段的基本要求，本教材共设计 2 册，每册有 8 个单元，分别供两个学年使用，各部分具体内容如下。

1. Pre-Reading Activities

此部分由视听材料和听说练习引出主题，激发学生对单元主题的兴趣及思考，有效促进学生对单元主题内容的理解。

2. In-Depth Reading

此部分为单元主体内容，由 Text A 和 Text B 两个部分组成。两篇文章围绕单元主题，各包含一篇课文和相关练习。单元主题涉及纺织技术、产业发展、服饰历史与文化，时尚设计等。文章选材既扎根本土文化又放眼世界，符合纺织服装专业学生的专业和兴趣，从价值引领、技能培养，以及开阔视野等多个维度，加强学生跨文化思辨意识，满足学生思想、情感与学习的需求。

3. Exercises

针对 Text A 和 Text B 设计的语言技能练习由以下内容构成。

（1）Reading Comprehension：包括课文理解和思辨问题，考查学生对课文主旨、重要细节、文章内涵的理解，同时也通过开放式问题及训练，培养学生的思辨与创新能力。

（2）Language Enhancement：旨在通过词汇练习加强学生对相关专业词汇的理解和运用，帮助学生学以致用，将语言学习和纺织服装专业知识有效结合起来。

针对 Text A，单独设计了以下练习。

（1）Translation：包含英译汉与汉译英两篇段落翻译，内容围绕单元主题，将专业内容进行拓展，在训练学生翻译能力的同时，引导学生理解和表达中西方在纺织服装产业方面存在的差异，提高学生的专业跨文化交际能力。

（2）Writing：写作内容从段落写作逐级进阶到篇章写作，从结构、方法和润色等方面培养学生的写作能力。

另外，针对 Text B，单独设计了 Sentence Structure 练习，该题主要针对文中出现的重点句型，通过补全翻译和句子改写的练习，帮助学生掌握语法使用规范，进一步夯实语言基础。

4. Extensive Reading

该部分文章内容同样与单元主题相关，学生通过自主阅读，训练英语阅读能力，进一步拓宽专业视野。

　　本系列教材总主编是西安工程大学胡伟华教授。第二册主编为郑卉蓉，参加编写的主要人员有石发德、李伟、廖燕、魏庆。本书的编写也得到西安工程大学人文社会科学学院的大力支持，在此表示由衷感谢！同时对本书参考文献的著作者、出版社编辑和各位工作人员一并表示诚挚谢意！

　　由于时间仓促，加之编者水平有限，不足之处敬请读者批评指正。

编者
2022 年 4 月

目 录
Contents

Unit 1

Domestic Textile and Garment Industry

A man who is not luxuriously dressed, but who is dignified, serious and virtuous, is awe-inspiring.

—Francis Bacon

For a wise and sensible person, the first requirement of clothing should always be decent and neat.

—George Washington

Pre-Reading Activities

1. Listen to the recording and answer the following questions.

(1) How much do China's textile and clothing exports account for of the world's total?

(2) What provides a strong purchasing power for the growth of textile consumption in China?

2. Listen to the recording again and fill in the blanks based on what you hear.

The output of China's textile industry has (1) _____ half of the world's total output, and China's textile and clothing exports (2) _____ 35% of the world's total. China is the world's largest textile producer and exporter, and the textile industry is not only the (3) _____ industry of the national economy, but also an industry with obvious international (4) _____ advantages. China's huge textile industry market provides a strong guarantee for the development of textile (5) _____ industry. The demand in the downstream market is relatively (6) _____ and the demand for medium- and high-grade textiles will continue to grow. From the perspective of (7) _____ power, the (8) _____ growth of Chinese residents' income provides a strong purchasing power for the growth of textile consumption in China.

3. Discuss the following questions with your partner.

(1) China is the world's largest textile producer and exporter, please share more about this with your partners.

(2) What is purchasing power in your opinion? How does it influence the textile consumption?

A Brief History of Chinese Textile Development

1   The narrow concept of textile refers to spinning and weaving. Spinning is the process from fiber to yarn. Weaving is the process from yarn to piece of cloth. The broad concept of textile relates to the whole process from raw fiber to final product.

2   The term *Textile* is a Latin word **originated from** the word *texere* which means to weave. Textile refers to a flexible material **comprising** of a network of natural or **artificial** fibers, known as yarn. Textiles are formed by weaving, knitting, **crocheting**, knotting and pressing fibers together. Textiles are fabrics or cloths, some of which may be woven and are used in the production of clothing and other textile-based goods such as blankets and rugs.

3   The history of textile is almost as old as that of human civilization. As time moves on, the history of textile has further enriched itself. In the 6th and 7th century BCE, the oldest recorded indication of using fiber came with the invention of flax and wool fabric at the **excavation** of Swiss lake inhabitants. In India, the culture of silk was introduced from China in about 140 BCE, while spinning of cotton **traces back to** 3000 BCE. In China, the discovery and consequent development of **sericulture** and spin silk methods **got initiated at** 2640 BCE, while in Egypt the art of spinning linen and weaving developed in 3400 BCE. The discovery of machines and their widespread application in processing natural fibers was a direct outcome of the industrial revolution of the 18th and 19th centuries. The discoveries of various synthetic fibers like nylon created a wider market for textile products and gradually led to the invention of new and improved sources of natural fiber.

4   The origin of Chinese textile is said to have been started by Leizu who raised silkworms and made silk, while bone needles have been found in the **archaeological** sites of **paleolithic** caveman.

5   In the **primitive** society, people gathered wild **kudzu**, hemp, silk, and so on. They used the feather of the birds and fur of beasts to weave into clothes through rubbing, tying and weaving. In the later period of primitive society, people learned the methods of manual production of textile materials, such as planting hemp, taking wool from wool breeding silkworm, and used many tools.

6   All processes at this stage are directly completed by human hands, hence it is called primitive textile. In the Shang and Zhou dynasties, the technology of silk weaving was greatly developed. In the Spring and Autumn Period and the Warring States Period, textile tools **evolved into** primitive manual textile machines such as reeling, spinning wheels and looms. The handwork became increasingly **exquisite**, and textiles were traded in large quantities, sometimes used as money. By the Spring and Autumn Period and the Warring States Period, the silk fabric was very exquisite. **Diversified** fabrics with rich colors make silk fabrics into noble materials. This is the stage of manual machine weaving from budding to forming.

7   From the Qin and Han dynasties to the end of the Qing dynasty, manual textile machinery gradually developed and improved, and a variety of forms appeared: reeling, spinning wheels by hand from a single **ingot** to a variety of complex ingot **pedal**. The loom has developed into two kinds: the plain machine and the flower machine.

8   Since the Song dynasty, the spinning wheel appeared to adapt to the production of **collectivized** workshops. Spinning, weaving, dyeing and finishing techniques became more and more mature, and today's main fabrics (plain weave, twill, satin) have appeared in the Southern Song dynasty. This is the stage of development of manual machine weaving.

9   By the 18th century, Western Europe developed the power machine textile,

and lots of foreign yarn and foreign cloth produced by machine were **dumped into** China, which influenced China's manual textile industry.

10   After the defeat of the Opium Wars, European textile technology was introduced in 1870. Large textile mills were **set up** but developed slowly. This is the formation stage of large-scale industrial textile.

11   After the founding of the People's Republic of China, the state **implemented** the policy of developing the textile industry in order to solve the problem of people's clothing as soon as possible. Consequently, China's textile industry has entered an unprecedented period of development.

12   Since 1970, China has **put forward** the policy of developing natural and chemical fiber, and gradually shifted the focus of textile industry to chemical fiber industry which has greatly developed.

13   Now the textile industry has been different, with the continuous development of the society, the textile production industry is more intelligent and **high-end**. The product becomes various, functional, and **superior**. Textile products are not only used in clothing and household, but also in medical, construction, **aerospace**, military defense, food and other fields.

Notes

Leizu   嫘祖，又名累祖，《山海经》中写作"雷祖"。中国远古时期人物，为西陵氏之女，轩辕黄帝的元妃。她发明了养蚕，史称嫘祖始蚕。

texere   拉丁文，有"编织"之义。

Yin Ruins   殷墟，位于河南安阳。

New words and phrases

comprise / kəm'praɪz /	v. ① to have sb/sth as parts or members 包括；包含 ② to be the parts or members that form sth 形成；组成

artificial /ˌɑːtɪˈfɪʃl/	*a.* ① made or produced to copy sth natural; not real 人工的；人造的；假的 ② created by people; not happening naturally 人为的；非自然的 ③ not what it appears to be 虚假的；假装的
crochet /krəʊˈʃei/	*v.* a way of making clothes, etc. from wool or cotton using a special thick needle with a hook at the end to make a pattern of connected threads 钩针编织；钩编工艺
excavation /ˌekskəˈveɪʃn/	*n.* ① [C, U] the activity of digging in the ground to look for old buildings or objects that have been buried for a long time（对古物的）发掘；挖掘 ② [C, usually pl.] a place where people are digging to look for old buildings or objects 发掘现场 ③ [U] the act of digging, especially with a machine 挖掘；开凿；挖土
sericulture /ˈserɪˌkʌltʃə/	*n.* the rearing of silkworms for the production of raw silk 养蚕
archaeological /ˌɑːkɪəˈlɒdʒɪkl/	*a.* related to or dealing with or devoted to archaeology 考古学的；考古学上的
paleolithic /ˌpælɪəʊˈlɪθɪk/	*a.* of or relating to the earliest period of the Stone Age characterized by rough or chipped stone implements 旧石器时代的
primitive /ˈprɪmətɪv/	*a.* ① [usually before noun] belonging to a very simple society with no industry, etc. 原始的；远古的 ② [usually before noun] belonging to an early stage in the development of humans or animals 原始的；人类或动物发展早期的 ③ very simple and old-fashioned, especially when sth is also not convenient and comfortable 发展水平低的；落后的 ④ [usually before noun] (of a feeling or a desire 感觉或欲望) very strong and not based on reason, as if from the earliest period of human life 原始本能的

kudzu /ˈkʊdzuː/	*n.* a kind of wild plant 野葛
exquisite /ɪkˈskwɪzɪt/	*a.* ① extremely beautiful or carefully made 精美的；精致的 ② (*formal*) (of a feeling 感觉) strongly felt 剧烈的；强烈的 ③ (*formal*) delicate and sensitive 微妙的；雅致的；敏锐的；敏感的
diversified /daɪˈvəːsɪfaɪd/	*a.* having variety of character or form or components 多样化的
ingot /ˈɪŋgət/	*n.* a solid piece of metal, especially gold or silver, usually shaped like a brick （尤指金、银的）铸块；锭
pedal /ˈpedl/	*n.* a flat bar on a machine that you push down with your foot in order to make parts of the machine move or work 踩踏板
collectivized /kəˈlektɪvaɪzd/	*v.* characterized by the principle of ownership by the state or the people of the means of production 使集体化
implement /ˈɪmplɪment/	*v.* put in force 实施；执行
high-end /ˌhaɪˈend/	*a.* expensive and of high quality 高档的；高端的；价高质优的
superior /suːˈpɪərɪə(r)/	*a.* ① better in quality than sb/sth else; greater than sb/sth else （在品质上）更好的；占优势；更胜一筹 ② higher in rank, importance or position （在级别、重要性或职位上）更高的
	n. a person of higher rank, status or position 级别（或地位、职位）更高的人；上级；上司
aerospace /ˈeərəʊspeɪs/	*n.* [U] the industry of building aircraft and vehicles and equipment to be sent into space 航空航天（工业）；航空航天技术
originate from	date from; derive from 源自；起源于
trace back to	retrospect to, be traced to 追溯到
get/be initiated at	begin, launch 开始；启动；发动

evolve into	to develop gradually, especially from a simple to a more complicated form 进化为；演变成
dump into	① to get rid of sth you do not want, especially in a place which is not suitable 把……丢弃，倾倒 ② to get rid of goods by selling them at a very low price, often in another country 向……倾销
set up	make ready or suitable or equip in advance for a particular purpose or for some use, event, etc. 建立；创建；配备
put forward	put in, bring forward 提出；拿出

▪ Reading Comprehension

Understanding the text

Answer the following questions.

1. How are textiles formed?

2. Why is the history of textile almost as old as that of human civilization?

3. When did the discovery and consequent development of sericulture and spin silk methods begin in China?

4. What is the earliest known origin of textile?

5. What does primitive textile refer to?

6. What is the result of manual textile machinery that was gradually developed and improved from the Qin and Han dynasties to the end of the Qing dynasty?

7. What influenced China's manual textile industry?

8. What is the formation stage of large-scale industrial textile?

9. According to your understanding, what is smart textile compared to traditional one?

10. What other fields can textile products be used besides clothing and household?

Critical thinking

I. Make a Presentation.

Please explain the historical context of the Chinese proverb *You can't buy all the Songjiang cloth, you can't get all the Weitang yarn.* (买不尽松江布，收不尽魏塘纱。) Work in pairs and try to prepare for a presentation in class.

II. Work in pairs and discuss the following questions.

1. Which dynasty's proverb is it?
2. What does "Songjiang cloth" specifically refer to?
3. How do you understand "The tax policy of government encouraged farmers to shift from growing rice to planting cotton"?

◤ Language Enhancement

Words in use

Fill in the blanks with the words given below. Change the form when necessary. Each word can be used only once.

comprise	artificial	crocheting	excavation	archaeological
primitive	exquisite	collectivize	implement	aerospace

1. The juice contains no _____ preservatives.
2. Most large businesses were _____ at the start of the war.
3. The company produces components for the _____ industry.
4. The committee is _____ of eight members.
5. This fossil is the skull of a _____ person.
6. This _____ embroidery won people's great admiration.
7. The crafts teacher is skillful in knitting, _____, embroidery, and the use of the hand loom.
8. All government institutions functionaries must _____ state laws, decrees and policies.

9. As an important _____ city, Xi'an has unearthed many important relics in recent years.

10. The _____ of the buried city took a long time.

Banked cloze

Fill in the blanks by selecting suitable words from the word bank. You may not use any of the words more than once.

A. interweaving	F. extraordinary	K. craftsmanship
B. heritage	G. primitive	L. involved
C. global	H. prosperity	M. visibly
D. associated	I. symbolically	N. spontaneous
E. excellent	J. elaborate	O. integral

From the earliest times, man has clothed himself in fabrics of one sort or another. At first, we used animal skins and furs directly as clothing. Then we learned to make crude materials by 1. _____ flexible strands, such as vines and creepers, into 2. _____ forms of cloth. This was the beginning of textile manufacture, which has now grown into one of the world's great industries.

Chinese textiles enjoy an excellent 3. _____ in the textile sector and occupy a prominent position in the global textile market. Chinese textiles are world famous and 4. _____ for their fine quality and profound symbolic meanings. Textiles in China often form an 5. _____ aspect of its heritage and 6. _____ reflect its tradition and culture.

In China, the textile is often closely associated with 7. _____ and involved in the process of 8. _____ rituals. Parents' 9. _____ love for their children is most visibly reflected through the excellent clothes they provide on festive occasions for their children. These clothes are made up of expensive materials and excellent 10. _____.

Expressions in use

Fill in the blanks with the expressions given below. Change the form when necessary. Each expression can be used only once.

originate from	trace back to	get/be initiated at
put forward	be assigned by	evolved into
dump into	set up	

1. All theories _____ practice and in turn serve practice.
2. The effluent from the factory was _____ the river.
3. Through reading, I can _____ ancient Egypt, one of the places where human civilization originated.
4. In fact, in many professions, important business transactions _____ a social event.
5. This contract may not _____ Seller without Buyer's written consent.
6. The pandemic _____ a global crisis since the beginning of the year.
7. They _____ many valid reasons for not exporting.
8. The city police _____ roadblocks to check passing vehicles.

■ Translation

I. Translate the following paragraph into Chinese.

In the beginning, clothing would have been very basic: it would have come from the skins and fur of animals. At some point, instead of wrapping oneself in one single large skin, humans began to stitch pieces of skin together. The earliest clothes in ancient society would have been made from plants, such as flax which can be converted into a textile called linen and cotton plants for cotton cloth. The domestication of animals added clothing made from wool textiles. Feathers, shells, leaves, and dyes were used to ornament clothing and create patterns, designs, and figures.

II. Translate the following paragraph into English.

最初用纱织成的布粗糙不匀。经过几百年的实践，纺纱技艺有了改进。除了过去用来纺纱的动物纤维及韧皮纤维外，人们开始使用更细的纤维，比如棉花。棉花逐渐成为纺织纤维中最重要的材料之一，用棉花纺织的东西价廉耐用。除棉花外，人们也用羊毛来弥补棉花的不足，羊毛织品漂亮、暖和、质优。

Paragraph Writing

How to develop a descriptive essay

Description is to paint in the mind of the reader an image or a series of images by means of language. A descriptive essay is a type of essay that provides a vivid description of person, place, scene, object, experience, memory, etc. in the way the reader can easily get its physical realization. More than many other types of essays, descriptive essays strive to create a deeply involved and vivid experience for the reader. Great descriptive essays achieve this affect not through facts and statistics but by using detailed observations and descriptions. We should always keep in mind that the essence of a descriptive essay is to make the readers see, hear, taste, smell or feel what we are writing about.

The purpose of a purely descriptive essay is to:
(1) involve the reader enough so he or she can actually visualize the things being described;
(2) make narration or exposition more lively, colorful and exciting when combined with them.

Text A of this unit includes many vivid descriptions. Let's take paragraph 3 for example, which aims to describe the history of textile in time order.

The history of textile is almost as old as that of human civilization. As time moves on, the history of textile has further enriched itself. In the 6th and 7th century

BCE, the oldest recorded indication of using fiber came with the invention of flax and wool fabric at the excavation of Swiss lake inhabitants. In India, the culture of silk was introduced in about 140 BCE, while spinning of cotton traces back to 3000 BCE. In China, the discovery and consequent development of sericulture and spin silk methods got initiated at 2640 BCE, while in Egypt the art of spinning linen and weaving developed in 3400 BCE. The discovery of machines and their widespread application in processing natural fibers was a direct outcome of the industrial revolution of the 18th and 19th centuries. The discoveries of various synthetic fibers like nylon created a wider market for textile products and gradually led to the invention of new and improved sources of natural fiber.

How can we make the readers experience the same sensory and emotional effects we are trying to communicate? It is by no means an easy job. Here are the steps we can follow when writing a descriptive essay.

First, we should have an introduction to put forward our thesis statement which states our dominant impression about a subject. Here are some examples:

The pet shop was noisy.

The restaurant was crowded.

The bus terminal was frightening.

The locker room was in an uproar.

Then, in the main body, we have the supporting details stated in separate paragraphs. Careful selection of details is the key to successful description. Once we have selected the most powerful details, we should arrange them in order. We can put them in the order of importance, time and so on. We should also use as many senses as possible when describing a scene. Chiefly we will use sight, but to some extent we may be able to use touch, hearing, smell, and perhaps even taste. Remember that it is through the richness of our sensory impressions that the reader will gain a clear picture of the scene.

The last part is the conclusion, where we should draw together all of the details we give to provide a final impression, which should help reinforce the attitude revealed in the thesis statement.

Structured writing

Read the sample essay and see how the description is developed.

Topic
Huntington Beach

Introduction
Thesis statement: The place where I feel most comfortable is my hometown of Huntington Beach.

Body
Detail 1: The sights of Huntington Beach make me relax and calm.
Detail 2: The sounds of Huntington Beach are in perfect harmony.
Detail 3: Everything on Huntington Beach has its own unique feel.

Conclusion
I find Huntington Beach most comfortable because of its sights, sounds, and its unique feel.

Sample essay
The place where I feel most comfortable is my hometown of Huntington Beach. Huntington Beach is in California about 40 miles south of Los Angeles. It is known as the surfing capital of the world and the best beach in Orange County.

I relax as I watch the surfers gently glide over the tumbling ocean waves. The swaying palm trees and the rolling sand dunes calm me down. The pier, as red as ruby, stands out to me among the deep blue ocean. As I lie on the sand, I see kites rise above the clouds and soar gracefully in the wind. The fishermen cast their lines off the pier hoping to feel tug and reel in big fish.

The sounds of Huntington Beach are in perfect harmony. Seagulls squawk as they soar overhead searching for food. The howling wind whistles through the beach like an arriving train. As the crash of the waves thunders through my ears, it brings me back to reality while I daydream. The sound of the oil rigs across Main Street is like a steady drum keeping the beat in an orchestra of sounds.

Everything on Huntington Beach has its own unique feel. The salty air blowing on my face feels wet and cool as it passes by. The feeling of the grainy sand is comforting to my feet as I walk across the shore. When

I plunge into the ocean's salty water, it feels refreshing to my skin, like a glass of water on a hot day.

I find Huntington Beach most comfortable because of its sights, sounds, and its unique feel mentioned in this essay. When I'm there, I feel totally relaxed as if I were in my own little world.

Write an essay of no less than 200 words on one of the following topics. One topic has an outline that you can follow.

Topic

Making a public speech

Introduction

Thesis statement: I had never thought I could speak in public, but I finally conquered my fear of it.

Body

Detail 1: I practiced my speech a lot with written notes for my English class.
Detail 2: I picked up the notes for my biology test instead of speech notes that day, but I made it.

Conclusion

The three-minute speech that day made me realize that people can always overcome their odds if they are motivated.

More topics

Topic 1: My favorite restaurant
Topic 2: Our college campus

Market Prospect of China's Textile and Garment Industry

1   China is a big textile country, and the textile industry is one of the **pillar** industries of the national economy. In 2018, the total amount of main business of China's textile **enterprises** above scale reached 5.37 **trillion** yuan, an increase of 2.9 percent over the same period last year. In 2019, the development of China's textile industry is facing a marked increase in domestic and foreign risks and is facing multiple pressures such as a more complex trade environment, a multiple demand in domestic and foreign market and a continuous rise in **comprehensive** costs. Therefore, the driving force of investment growth of textile enterprises is weak, the level of efficiency is declining, and the main economic operation **indicators** are **slowing down**. In 2019, the operating income of China's textile and clothing enterprises above scale reached 4.94 trillion yuan, down 1.47 percent **compared with** the same period last year.

2   According to the data of the National Bureau of Statistics of China, in 2019, the retail sales of clothing, shoes, hats, **knitwear** and textiles above the national limit was 1.35 trillion yuan, an increase of 2.9 percent over the same period last year, slowing down 5.1 percent from 2018. Retail sales of online wearable goods nationwide increased by 15.4 percent **year-on-year**, continuing to maintain a good **double-digit** growth level, but 6.6 percent lower than the **previous** year.

3   Since 2020, with the global spread of the pandemic, the industry is facing **tremendous** environmental pressure and **systemic** risks, both ends of supply and demand of the industry have experienced **unprecedented** shocks.

4   In the past 20 years, the climate index of textile enterprises has changed **in a fluctuating manner, reaching a peak** in 2003, 2007, 2011, 2013 and 2017, and a **phased trough** in 2002, 2004, 2009, 2012 and 2016. In 2013 and 2014, the textile industry climate index remained in the growth range as a whole, but the growth trend has slowed down compared with the previous year. In recent years, the climate index of the textile industry is basically the same as that of the previous year.

5   **To sum up,** the domestic economic **deceleration** shift presents a new **normal** from high-speed growth to medium-to-high-speed growth, and the growth rate of China's textile industry **slackens** obviously. As the textile machinery industry in the upper reaches of the textile industry chain, it **is affected by** the domestic market demand **to a certain extent**. According to the *Textile Industry Development Plan (2016–2020)*, China's economic development has entered a new normal during the 13th Five-Year Plan period. Although the development of environment and situation of the textile industry are **undergoing profound** changes, the **overall** development opportunities **outweigh** the challenges. If we **make good use of** the **strategic** opportunities of the new round of science and technology and industrial change, the textile industry will maintain medium- and high-speed development, and the average annual growth rate of industrial added value of textile enterprises above scale will be maintained at 6% and 7%. This provides policy support for the development of textile **machinery** industry.

6   The growth of the textile industry is driven by many factors and is **sustainable** for the following reasons:

1) The national policy **is favorable to** promote the development of the textile industry. In recent years, the state has issued a series of policies to support the **transformation** and **upgrading** of the textile industry. In 2016, Ministry of Industry and Information Technology of the People's Republic of China issued the *Textile Industry Development Plan (2016–2020)*. As a special plan to guide the development of the

textile industry during the 13th Five-Year Plan period, it will promote the transformation and upgrading of the textile industry and create new **competitive** advantages. With the support of the national policy, the industrial structure of the textile industry has been continuously **optimized**, and the technological level has been continuously enhanced. China's textile industry is changing from traditional and investment-driven production factors to innovation-driven production factors, from traditional labor-intensive industries to science and technology industries.

2) "Internet +" has become a new marketing mode in textile industry. "Internet + Textile" is a new development point in the textile industry. The popularity of the Internet has led to the rapid development of e-commerce. Textile and clothing online retail continues to maintain rapid growth. In 2018, retail sales of online wearable goods increased by 22% compared with the same period last year, an increase of 1.7 percentage points over the previous year. Tmall's "double 11" data show that the clothing **category accounts for** 27% of the brands with a **turnover** of more than 100 million, ranking first and **exceeding** the second category by 6 percentage points. The **construction** of brand online channels in the textile and clothing industry has become increasingly **mature**, and the intelligent supply chain system and intelligent manufacturing capability have been steadily improved.

3) **Regional** industrial structure adjustment, the Silk Road Economic Belt and the 21st-century Maritime Silk Road brings new opportunities. In recent years, the **layout** of China's textile industry has changed obviously, and one of its **remarkable** features is industrial **transfer**. China's textile industry presents a transfer path from east to west and from inside to outside. Textile and clothing enterprises actively implement the Belt and Road Initiative, and **carry out** layout in countries along the route, and begin to **extend** to Central and Eastern Europe and Africa, making full use of the superior resources of relevant countries to carry out **vertical** industrial chain

layout and enhance the international competitiveness of enterprises.

4) Scientific and technological innovation promotes industrial intelligent production. The textile industry actively uses intelligent manufacturing as a means to open up the data flow of the whole textile industry chain and key links of production and manufacturing, **enforce** the reform of production mode and business model, greatly improve the labor productivity and production **flexibility** of the textile industry, product quality and energy consumption of resources, and promote the development of China's textile industry to a high-end direction.

7　　To sum up, the development of domestic textile industry is driven by many factors, such as national policy, Internet development, regional industrial structure adjustment and intelligent production, and the growth is sustainable to a certain extent.

Notes

National Bureau of Statistics of China　国家统计局

the 13th Five-Year Plan　第十三个五年计划

supply chain system　供应链体系

labor-intensive industries　劳动密集性产业

Central and Eastern Europe　中东欧

New words and phrases

pillar /ˈpɪlə(r)/

n. ① a tall vertical cylindrical structure standing upright and used to support a bridge, the roof of a building, etc., especially when it is also decorative 柱子；桥墩 (尤指兼作装饰的) ② a large round stone, metal or wooden post that is built to remind people of a famous person or event 纪念柱

enterprise /ˈentəpraɪz/   *n.* ① a company or business 公司；企业单位；事业单位 ② the development of businesses by the people of a country rather than by the government 企业发展；企业经营；企业活动

trillion /ˈtrɪljən/   *n.* ① one million million 万亿；兆 ② a trillion (or) trillions 大量；无数

comprehensive /ˌkɒmprɪˈhensɪv/   *a.* ① including all, or almost all, the items, details, facts, information, etc., that may be concerned 全部的；所有的；（几乎）无所不包的；详尽的 ② (BrE) designed for students of all abilities in the same school 综合性的（接收各种资质的学生）

indicator /ˈɪndɪkeɪtə(r)/   *n.* ① a sign that shows you what sth is like or how a situation is changing 指示信号；标志；迹象 ② a device on a machine that shows speed, pressure, etc. 指示器；指针

knitwear /ˈnɪtweə(r)/   *n.* knitted clothing 针织衫；针织衣物

double-digit /ˌdʌbl ˈdɪdʒɪt/   *a.* number that is between 10 and 99 两位数的

previous /ˈpriːvɪəs/   *a.* immediately before the time you are talking about（时间上）稍前的

tremendous /trəˈmendəs/   *a.* ① extraordinarily large in size, extent, amount, power or degree 巨大的；极大的；② extraordinarily good or great 极好的；精彩的；了不起的

systemic /sɪˈstiːmɪk/   *a.* affecting an entire system 系统的；影响全面的

unprecedented /ʌnˈpresɪdentɪd/   *a.* having no precedent 前所未有的；空前的；没有先例的

fluctuating /ˈflʌktʃueɪtɪŋ/   *a.* having unpredictable ups and downs 波动的；起伏不定的

phased /feɪzd/   *a.* arranging to do sth gradually in stages over a period of time

trough /trɒf/   *n.* ① a long narrow container from which farm animals drink or eat 水槽 ② a low point in a process that has regular high and low points 低谷

deceleration /ˌdiːseləˈreɪʃn/   *n.* a decrease in rate of change 减速（度）；降速

normal /ˈnɔːml/   *n.* sth regarded as a normative example 标准，准则

slacken /ˈslækən/   *v.* to become or to make sth become slow or slower （使）变缓，松弛

undergo /ˌʌndəˈɡəʊ/   *v.* to experience (change, unhappiness, etc.) 经历；经受（变化，不快的事等）

profound /prəˈfaʊnd/   *a.* very great; felt or experienced very strongly 巨大的；深切的；深远的

overall /ˈəʊvərɔːl/   *a.* including all the things or people that are involved in a particular situation; general 全面的；综合的；总体的

outweigh /ˌaʊtˈweɪ/   *v.* to be heavier than sth 重于；大于；超过

strategic /strəˈtiːdʒɪk/   *a.* ① highly important to or as an integral part of a strategy or plan of action, especially in war 具有战略意义的；重要的；合适的 ② relating to or concerned with strategy 战略性的；战略上的

machinery /məˈʃiːnəri/   *n.* ① machines as a group, especially large ones（统称）机器；（尤指）大型机器 ② a system of means and activities whereby a social institution functions 组织；机构；系统；体制

sustainable /səˈsteɪnəbl/   *a.* ① involving the use of natural products and energy in a way that does not harm the environment 不破坏生态平衡的；合理利用的 ② that can continue or be continued for a long time 可持续的

transformation /ˌtrænsfəˈmeɪʃn/   *n.* a complete change in sb/sth（彻底的）变化，改观，转变，改革

upgrade /ˌʌpˈɡreɪd/   *v.* ① to make a piece of machinery, system, etc. more powerful or efficient 升级；提高；改进；提升 ② to give sb a more important job 提拔；提高待遇；优待

competitive /kəmˈpetətɪv/   *a.* ① used to describe a situation in which people or organizations compete against each other 竞争的 ② as good as or better than others（与……）一样好的；（比……）更好的；有竞争力的

optimize /ˈɒptɪmaɪz/   *v.* make full use of 使最优化；充分利用

category /ˈkætəɡəri/   *n.* a group of people or things with particular features in common（人或事物的）类别；种类

turnover /ˈtɜːnəʊvə(r)/   *n.* ① the total amount of goods or services sold by a company during a particular period of time（一定时期内的）营业额，成交量 ② the rate at which goods are sold in a store and replaced by others（商店的）货物周转率，销售比率

exceed /ɪkˈsiːd/   *v.* far beyond what is usual in magnitude or degree 超过，超出（某数量、限制等）

construction /kənˈstrʌkʃn/   *n.* [U] the process or method of building or making sth, especially roads, buildings, bridges, etc. 建筑；建造；施工

mature /məˈtʃʊə(r)/   *a.* ① (of a person, a tree, a bird or an animal 人、树木、鸟或动物) fully grown and developed 成熟的；发育完全的 ② (of a child or young person 儿童或年轻人) behaving in a sensible way, like an adult 明白事理的；成熟的；像成人似的

v. to become fully grown or developed 成熟；长成

regional /ˈriːdʒənl/	*a.* of or relating to a region 地区的；区域的；地方的
layout /ˈleɪaʊt/	*n.* the way in which the parts of it are arranged 布置；设计；布局
remarkable /rɪˈmɑːkəbl/	*a.* unusual or surprising in a way that causes people to take notice 非凡的；奇异的；显著的；引人注目的
transfer /trænsˈfɜː(r)/	*v.* ① to move from one place to another; to move sth/sb from one place to another（使）转移，搬迁 ② to move from one job, school, situation, etc. to another; to arrange for sb to move（使）调动；转职；转学；改变（环境）③ to officially arrange for sth to belong to sb else or for sb else to control sth 转让；让与
extend /ɪkˈstend/	*v.* ① to make sth longer or larger 使伸长；扩大；扩展 ② to make sth last longer 延长；使延期
vertical /ˈvɜːtɪkl/	*a.* ① going straight up or down from a level surface or from top to bottom in a picture, etc. 竖的；垂直的；直立的 ② having a structure in which there are top, middle and bottom levels 纵向的
enforce /ɪnˈfɔːs/	*v.* ① to make sure that people obey a particular law or rule 强制执行，强行实施（法律或规定）② to make sth happen or force sb to do sth 强迫；迫使
flexibility /ˌfleksəˈbɪləti/	*n.* the quality of being adaptable or variable 灵活性；适应性
slow down	become slow or slower; become less tense, rest, or take one's ease 放慢；减速；放松
compare with	comparable, in contrast with 比得上；与……相比

year-on-year	at the same point each year when compared with the previous year 同比增长；与上年同期数字相比
in a … manner	in a sense 在某种意义上
reach a peak	culminate 达到顶峰
be affected by	be influenced by 受到……的影响
to sum up	to conclude 总结
to a certain extent	to a certain degree 在一定程度上
make good use of	use; utilize; take advantage of 利用
be favorable to	be instrumental in 有利于
account for	to make up or form (a part of sth), to be the reason or explanation for 占比，解释；说明
carry out	perform, accomplish, implement, fulfill, come true 执行；实行；贯彻；实现；完成

▪ Reading Comprehension

Understanding the text

Choose the correct answer in each of the following questions.

1. What is the total amount of the operating income of China's textile and clothing enterprises above scale in 2019?

    A. 5.37 trillion yuan.

    B. 4.94 trillion yuan.

    C. 1.35 trillion yuan.

    D. It's unknown.

2. The textile industry is facing tremendous environmental pressure and systemic risks in 2020 _____.

    A. because of the financial crisis

    B. because of the global spread of the pandemic

    C. because the textile industry is not important any more

D. because the policy has greatly changed

3. The textile industry climate index remain in the growth range as a whole in _____.

A. 2011 and 2012

B. 2013 and 2014

C. 2002 and 2004

D. 2009 and 2012

4. In 2018, retail sales of online wearable goods were _____ compared with the same period last year.

A. increased by 27%

B. decreased by 27%

C. decreased by 22%

D. increased by 22%

5. The development of domestic textile industry is driven by _____.

A. national policy

B. Internet development

C. regional industrial structure adjustment and intelligent production

D. all of above

Critical thinking

Work in pairs and discuss the following questions.

1. What does the Belt and Road Initiative bring to the textile industry?

2. Discuss the relationship between the textile industry and "Internet +".

3. Why is the growth of the textile industry sustainable?

Language Enhancement

Words in use

Fill in the blanks with the words given below. Change the form when necessary.
Each word can be used only once.

enterprise	undergoing	tremendous	unprecedented
upgrading	comprehensive	outweigh	fluctuating
optimizing	strategic		

1. Financial management is key in any company or _____.
2. There are much brighter prospects for a _____ settlement than before.
3. He plays the character with _____ concentration combined with a pleasing modesty.
4. We are now witnessing an _____ increase in violent crime.
5. People close to him said last month his health had been _____.
6. Several of the airline's planes are temporarily out of commission and _____ safety checks.
7. The advantages far _____ the disadvantages.
8. The problem of countryside is comprehensive and _____ ever since in this country.
9. The _____ and updating of digital products have accelerated for the past several years.
10. Doctors are concentrating on understanding the disease better, and on _____ the treatment of pneumonia.

Expressions in use

Fill in the blanks with the expressions given below. Change the form when necessary. Each expression can be used only once.

slow down	year-on-year	reach a peak	be favorable to
account for	compare with	to a certain extent	make good use of

1. We turned to see the approaching car _____.
2. This would be the biggest _____ fall for more than 60 years.
3. The flowers here do not _____ those at home.
4. By 2050, the aging population of this country will _____ total of 437 million, at that time, nearly one-third of the population is the elderly.
5. The social services _____ a substantial part of public spending.
6. It is suggested that appropriate reduction of temperature may _____ the treatment of the disease.
7. We must _____ the available space.
8. _____, the article reflected the reality.

Sentence structure

Complete the following sentences by translating the Chinese into English, using "with ... , sb/sth is facing ..." structure.

Model: _____（随着疫情在全球蔓延），
the industry _____（面临）tremendous environmental pressure and systemic risks.
→With the global spread of the pandemic, the industry is facing tremendous environmental pressure and systemic risks.

1. _____（随着就业市场竞争的日趋激烈），
more and more college students _____（面临）the problem of employment difficulties.
2. _____（随着各国之间交流和合作的加强），
the textile and clothing industry _____（面临）unprecedented

opportunities and challenges.

Rewrite the following sentences by using "the same as ..."

> **Model:** In recent years, the climate index of the textile industry basically equals to that of the previous year.
>
> →In recent years, the climate index of the textile industry <u>is</u> basically <u>the same as</u> that of the previous year.

1. Last year, the results of the online sales stores and offline physical stores of this product were the same.

   _____

   _____.

2. Generally speaking, swimming for one hour and running for five kilometers burn the same amount of calories.

   _____

   _____.

Extensive Reading

China to Up Its Textile Recycling Capability and Sustainable Development

1   China, producer of half the world's textile fiber, has unveiled a guideline that aims to significantly beef up its capability to recycle textile waste, most of which is non-biodegradable. Experts have lauded the initiative for its potential role in promoting low-carbon, circular economic development, saying it will contribute to the country's ambitious climate and pollution targets.

2   The country aims to recycle a quarter of its textile waste and use it to produce 2 million metric tons of recycled fiber annually by 2025, according

to a document unveiled by the National Development and Reform Commission, the country's top economic planner, and the ministries of industry and information technology, and commerce. Five years on from 2025, a relatively complete system for textile waste recycling will have been established in the country, it said. By then, China will be able to recycle 30 percent of its textile waste and produce 3 million tons of recycled fiber annually.

3   "As a key part of establishing and then improving a green, low-carbon and circular economic system, recycling helps in resource conservation and the reduction of pollutants and carbon emissions," the commission said in a news release. Zhao Kai, executive vice-president of the China Association of Circular Economy, said that as people live wealthier lifestyles, there is a greater demand for clothing, which creates more waste. Only about one-fifth of the roughly 22 million tons of textile waste generated in the country in 2020 was recycled. China produced only 1.5 million tons of recycled fiber that year. "There is a lot of room for improvement in the country's capability to recycle textile waste," Zhao stressed. The guideline will hopefully address a series of weak links in textile waste recycling, he continued. The document, for example, vows to introduce preferential policies to motivate companies to improve product design, so that they can be more easily dismantled, classified and recycled after being discarded. Aside from improving the network for collecting textile waste, he said, the guideline also aims to address problems that hinder reuse and recycling. The country will hammer out and strive to improve the industrial standards and norms on cleaning, disinfection, epidemic control and trading of secondhand clothes, he said. Export control of used garments will be further strengthened. The guideline pledged to foster leading enterprises in an endeavor to promote the use of recycled fiber in the textile, construction, automobile, agriculture and environmental protection sectors, he noted.

4   Other experts stressed the environmental benefits the document will bring. Tang Shijun, former head of the Quartermaster Equipment Research

Institute, said the use of every kilogram of recycled textile waste will help reduce carbon dioxide emissions by 3.6 kg and save 6,000 liters of water. Sun Huaibin, vice-president of the China National Textile and Apparel Council, said China currently gets recycled fiber as a raw material mainly from plastic bottles and industrial textiles. Only a very small amount of old clothes are recycled.

5   As the seventh largest economy in the world, China's fashion industry accounts for up to 40%. The "climate action power" of China's fashion industry is of immeasurable significance to the achievement of the 2030 UN sustainable development goals. Recently, Sun Ruizhe, the president of China National Textile and Apparel Council (CNTAC), led a total of 16 Chinese fashion industry climate action delegations (representing all stakeholders in the industry chain) to Madrid, Spain, to participate in the United Nations Framework Convention on Climate Change (UNFCCC) Twenty-Fifth Conference of the Parties (COP25). The delegation fully demonstrated the climate leadership of China's textile and apparel industry to the world in terms of climate innovation, climate empowerment, climate youth, sustainable fashion, sustainable consumption and other areas of practical innovation, as well as participation in the governance of global climate.

6   The responsibility background of global climate change has caused Chinese fashion companies to bear double pressures for a long time: on the one hand, they must respond to the fluctuations in the interests of policies and customer needs; on the other hand, they must adapt to new business rules and seek initiative for their development space. As the industry guidance and service organization, the Social Responsibility Office of CNTAC shall start with the ideology of the Chinese fashion industry, and through a series of professional training, innovation and practice to help enterprises from "passive" turned into "active". In recent years, China's textile and apparel companies have made amazing progress in areas such as energy saving and emission reduction. Sun Ruizhe, the president of

CNTAC, said in a speech at the "Enterprise Climate Action: Enabling and Innovation" conference that the fourth industrial revolution in the future will be the green industrial revolution plus artificial intelligence. Scientific and technological innovations such as energy and recycling materials solve the problem of sustainable development productivity.

7   Tackling climate change is not only a new opportunity for the global fashion industry, but also a direction for shared responsibility across industries. Global issues such as climate change, resource shortages, and population pressure are moving towards specialization and subdivision. For China's textile and apparel industry, which contributes to sustainable fashion, climate change will no longer be a burden and "cost", but a new business development opportunity. The performance of the Chinese fashion industry climate action delegation at COP25 is also a clear signal from China to the world that new perspectives and new models for global climate leadership will be in China.

Unit 2

Introduction to Textile Technology

To think that the new economy is over is like somebody in London in 1830 saying the entire industrial revolution is over because some textile manufacturers in Manchester went broke.

—*Alvin Toffler*

I guess the big thing is that I don't buy anything first-hand. It's a personal policy I have for all sorts of reasons. If you research to the textile industry yourself, you'll know why. I came to it personally.

—*Ezra Miller*

Pre-Reading Activities

1. Listen to the recording and answer the following questions.

(1) What is knitting mainly divided into?

(2) How does warp knitting lace expand with the development of the textile industry?

2. Listen to the recording again and fill in the blanks based on what you hear.

Warp (1) _____ lace is made by knitting on a warp knitting machine, also known as knitted lace or lace. It (2) _____ in Europe in the 16th century and is widely used in the decoration of (3) _____ clothing. With the development of the textile industry, warp knitting lace has (4) _____ from clothing decoration to home decoration and other fields, and with the (5) _____ of production technology and machinery, its products have been continuously (6) _____ and beautiful. It has (7) _____ the rapid development of warp knitting lace industry. There are many kinds of warp knitting lace, which are mainly (8) _____ according to the machines that produce lace.

3. Discuss the following questions with your partner.

(1) What are warp knitted fabrics often made of ?

(2) Please discuss the differences between warp knitting and weft knitting and try to illustrate them.

Principles and Classification of Spinning Technology

1   The basic principles of spinning can be divided into the following four types: removing impurities, releasing, loosing and carding.

2   The first one is removing impurities. Spinning is a science that studies the processing of textile staple fibers into yarn. Yarn is generally made of many staple fibers of different length by splicing and twisting of long continuous filaments. In the spinning process, we first need to remove miscellaneous defects, that is, the preliminary processing of raw materials, which is also known as the preparation of spinning materials. Since there are different kinds of raw materials, as well as different types and properties of impurities, processing methods and technic differ accordingly. The preliminary processing methods of raw materials are mainly physical methods (such as cotton ginning), chemical methods (such as hemp degumming, silk scouring) and a combination of physical and chemical methods (such as wool washing and peat removal).

3   The second one is releasing. The disorganized and transversely closely connected fibers are processed into a longitudinal sequence with smooth and clean yarns based on certain requirements, so it is necessary to turn the bulk fibers into a single fiber state, remove the horizontal connection of the fiber raw materials, and establish a strong longitudinal connection end to end. The former is called the release of fibers, and the latter the collection of fibers. The release of the fiber is to completely remove the transverse connection among the fibers, and the damage to the fiber must be reduced as much as possible. The collection of fibers is to make the loosened fibers re-establish an orderly longitudinal connection, which is

continuous. The fiber distribution in the set should be **uniform**, having a certain **linear density** and strength **at the same time**. Fiber **aggregation** needs to be added a certain twist. The collection process is not completed at one time but can only be done after many times of processing, such as carding, drafting, twisting and so on.

4   The third one is loosening. Opening loose is to **tear** large pieces of fiber **into** small pieces, even small fiber **bundles**. In a broad sense, hemp degumming is also a kind of loosening. With the progress of loosening, the bonding force between fibers and impurities is weakened, so that impurities are removed and the fibers are mixed at the same time. The loosening effect and the removal of impurities are not completed at one time, but gradually realized through the **rational configuration** of tearing, hitting and **segmentation**.

5   The fourth one is carding. The carding function is that the small pieces and bundles of fibers are further loosened into a single state by **a large number of** dense carding needles on the carding machine, **so as to** further improve the release of fibers. After carding, the transverse connection between the fibers is basically removed, and the effect of impurity removal and mixing is more sufficient. A large number of fibers are **curved** with **hooks**, and there is still a certain transverse connection between each fiber.

6   As for spinning technology, it is classified into ring spinning, compact spinning, rotor spinning and vortex spinning.

7   Ring spinning began in the 19th century and has been widely used at present. The spinning quality has reached a very high level, and the scope of application and variety **adaptability** have been expanded towards high-tech production technology. Although ring spinning has made **considerable** progress, the production technology is still not very ideal. The fiber width of the forming zone of ordinary ring spinning is obviously larger than that of the spinning **triangle**, indicating that some fibers on the edge of the triangle will be lost, or cannot be grasped by the yarn body,

resulting in uncontrolled fibers.

8   Compact spinning is divided into pneumatic compacting and mechanical compacting through compacting principle. Compact spinning technology can be used in cotton spinning, wool spinning, bast fiber spinning and chemical fiber spinning. In compact spinning, the fibers are compacted before spinning and thus optimally integrated into the yarn body. The compact yarn produced in this way convinces with a unique yarn structure and unique yarn properties. The flexibility of the compact-spinning machines and the compacting system is demonstrated in the versatile yarn variations and applications.

9   Rotor spinning has gradually become the most mature and widely used spinning method in the new type of spinning because of its short spinning process, strong spinnability and high speed. With more than 40 years of development, the number of rotor spinning equipment and the output of rotor spinning in China have accounted for one third of the world's total.

10   Vortex spinning is a new spinning method which uses fixed vortex spinning tube to replace the spinning cup with high-speed rotation. In the process of vortex spinning, fiber transfer, condensation, twisting and yarn formation are all completed with the help of air flow. The structure of the vortex spinning machine is simple, the high-speed rotary parts are eliminated, and the moving sliver is twisted with the help of the high-speed rotary air flow. As for the structural characteristics of vortex spinning yarn, vortex spinning is a kind of yarn with dual structure. The core fiber of the yarn is arranged in parallel and has no twist. Depending on the action of rotating air flow, the end fiber is wrapped and twisted on the outside of the core fiber to form the yarn.

11   Spinning technology will be further developed in the future, in which ring spinning, compact spinning, rotor spinning, eddy current spinning and staple fiber spinning systems will be developed rapidly.

Notes

cotton ginning 轧棉

ring spinning 环锭纺

compact spinning 紧密纺

rotor spinning 转杯纺

vortex spinning 涡流纺

New words and phrases

spin /spɪn/   *v.* ① ~ **(sth) (round/around)** to turn round and round quickly（使）快速旋转 ② to turn round quickly once（使）急转身；猛转回头；急转弯 ③ to make thread from wool, cotton, silk, etc. by twisting it 纺（线）；纺（纱）

remove /rɪˈmuːv/   *v.* ① to take sth/sb away from a place 移开；拿开；去掉 ② to take off clothing, etc. from the body 脱去（衣服等）；摘下 ③ to get rid of sth unpleasant, dirty, etc.; to make sth disappear 去除，排除（污渍、不愉快的事物等）；使消失 ④ to dismiss sb from their position or job 免除，解除（职务等）

impurity /imˈpjuəriti/   *n.* (*pl.*-ies) worthless or dangerous material that should be removed 杂质

release /rɪˈliːs/   *v.* ① to let sb/sth come out of a place where they have been kept or trapped 释放；放出；放走 ② to stop holding sth or stop it from being held so that it can move, fly, fall, etc. freely 放开；松开；使自由移动（或飞翔、降落等） ③ to express feelings such as anger or worry in order to get rid of them 发泄；宣泄

n. ① [U, sing.] the act of setting a person or an animal free; the state of being set free 释放；获释 ② [U, sing.] the act of making sth available to the public 公开；发行；发布 ③ [U, C] the act of letting a gas, chemical, etc. come out of the container where it has been safely held 排放；泄漏；渗漏 ④ [U, sing.] the feeling that you are free from pain, anxiety or some other unpleasant feeling 解脱；轻松感

carding /'kɑːdɪŋ/

n. to comb out and clean fibers of wool or cotton before spinning（纺织的）梳理

staple /'steɪpl/

a. ① forming a basic, large or important part of sth 主要的；基本的 ② relating to the textile fiber 纺织纤维的

n. ① a small piece of wire that is used in a device called a stapler and is pushed through pieces of paper and bent over at the ends in order to fasten the pieces of paper together 订书钉 ② a basic type of food that is used a lot 基本食物；主食 ③ sth that is produced by a country and is important for its economy 主要产品；支柱产品

v. to attach one thing to another using a staple or staples 用订书钉装订

yarn /jɑːn/

n. [U] thread that has been spun, used for knitting, making cloth, etc. 纱；纱线

splice /'splaɪs/

v. ① to join the ends of two pieces of rope by twisting them together 绞接，捻接（两段绳子）② to join the ends of two pieces of film, tape, etc. by sticking them together 胶接，粘接（胶片、磁带等）

n. the place where two pieces of film, tape, rope, etc. have been joined 胶接处；粘接处；绞接处

twist /twɪst/	v. to bend or turn sth into a particular shape 使弯曲，使扭曲（成一定形状）
filament /ˈfɪləmənt/	n. a very slender natural or synthetic fiber [植] 花丝；细丝
miscellaneous /ˌmɪsɪˈleɪnjəs/	a. consisting of a haphazard assortment of different kinds 各种各样的；混杂的
defect /ˈdiːfekt; dɪˈfekt/	n. a fault in sth or in the way it has been made which means that it is not perfect 缺点；缺陷；毛病
preliminary /prɪˈlɪmɪnəri/	a. happening before a more important action or event 预备性的；初步的；开始的
raw /rɔː/	a. ① not cooked 生的；未烹制的；未煮的 ② in its natural state; not yet changed, used or made into sth else 未经加工的；自然状态的
gin /ˈdʒɪn/	v. to separate the seeds from (cotton) with a cotton gin 轧花，轧棉
degum /diːˈɡʌm/	v. to free from gum 除胶
scour /ˈskaʊə(r)/	v. ① ~ sth (for sb/sth) to search a place or thing thoroughly in order to find sb/sth （彻底地）搜寻；搜查；翻找 ② ~ sth (out) to clean sth by rubbing its surface hard with rough material（用粗糙的物体）擦净；擦亮 ③ ~ sth (away/out)	~ sth (from/out of sth) to make a passage, hole, or mark in the ground, rocks, etc. as the result of movement, especially over a long period 冲刷成；冲刷出
combination /ˌkɒmbɪˈneɪʃn/	n. [U] the act of joining or mixing together two or more things to form a single unit 结合；联合；混合
peat /piːt/	n. a soft black or brown substance formed from decaying plants just under the surface of the ground, especially in cool wet areas 泥煤；泥炭

removal /rɪˈmuːvl/   *n.* the act of taking sth/sb away from a place 移除；去除；摘去

disorganized /dɪsˈɔːgənaɪzd/   *a.* badly planned; not able to plan or organize well 计划不周的；缺乏组织的；杂乱无章的

transversely /trænsˈvɜːsli/   *ad.* in a transverse manner 横着；横断地；横切地

longitudinal /ˌlɒŋgɪˈtjuːdɪnl/   *a.* ① going downwards rather than across 纵的；纵向的 ② connected with longitude 经度的

horizontal /ˌhɒrɪˈzɒntl/   *a.* flat and level; going across and parallel to the ground rather than going up and down 水平的；与地面平行的；横的

loosened /ˈluːsənd/   *a.* less strict or severe 疏松的；分散的

re-establish /riːɪˈstæblɪʃ/   *v.* to establish (sth) again 重建

uniform /ˈjuːnɪfɔːm/   *n.* the special set of clothes worn by all members of an organization or a group at work, or by children at school 制服；校服

     a. not varying; the same in all parts and at all times 一致的；统一的；一律的

linear /ˈlɪniə(r)/   *a.* of or in lines 线的；直线的；线状的

density /ˈdensəti/   *n.* [U] the quality of being dense; the degree to which sth is dense 密集；稠密；密度；浓度

aggregation /ˌægrɪˈgeɪʃn/   *n.* a group, body or mass composed of many distinct parts of individuals 聚集；集合体

bundle /ˈbʌndl/   *n.* a number of things tied or wrapped together; sth that is wrapped up （一）捆，包，扎

rational /ˈræʃnəl/   *a.* based on reason rather than emotions 合理的；理性的；明智的

configuration /kənˌfɪɡəˈreɪʃn/   *n.* ① an arrangement of the parts of sth or a group of things; the form or shape that this arrangement produces 布局；结构；构造；格局；形状 ② the equipment and programs that form a computer system and the way that these are set up to run（计算机的）配置

segmentation /ˌseɡmenˈteɪʃn/   *n.* the act of dividing sth into different parts; one of these parts 分割；划分；分割成的部分

curved /kɜːvd/   *a.* having a round shape 呈弯曲状的；弧形的

hook /hʊk/   *n.* a curved piece of metal, plastic or wire for hanging things on, catching fish with, etc. 钩；钓钩；挂钩；鱼钩

v. to fasten or hang sth on sth else using a hook; to be fastened or hanging in this way（使）钩住；挂住

adaptability /əˌdæptəˈbɪləti/   *n.* the ability to change (or be changed) to fit changed circumstances 适应性；可变性；适合性

considerable /kənˈsɪdərəbl/   *a.* great in amount, size, importance, etc. 相当多（或大、重要等）的

triangle /ˈtraɪæŋɡl/   *n.* a flat shape with three straight sides and three angles; a thing in the shape of a triangle 三角形；三角形物体

spinnability /spɪnəˈbɪlɪtɪ/   *n.* the quality of being suitable for spinning or the capability of being spun (used of textile fibers) 可纺性；纺丝性

replace /rɪˈpleɪs/   *v.* ① to be used instead of sth/sb else; to do sth instead of sb/sth else 代替；取代 ② to change sth that is old, damaged, etc. for a similar thing that is newer or better 更换；更新

rotation /rəʊˈteɪʃn/	n. ① [U] the action of an object moving in a circle around a central fixed point 旋转；转动 ② [U, C] the act of regularly changing the thing that is being used in a particular situation, or of changing the person who does a particular job 轮换；交替；换班
condensation /ˌkɒndenˈseɪʃn/	n. the process of a gas changing to a liquid（气体）冷凝，凝结
rotary /ˈrəʊtəri/	a. moving in a circle around a central fixed point 旋转的
dual /ˈdjuːəl/	a. having two parts or aspects 两部分的；双的
wrapped /ræpt/	a. covered with or as if with clothes or a wrap or cloak 有包装的
eddy /ˈedi/	n. a movement of air, dust or water in a circle（空气、灰尘或水的）旋涡；涡流
	v. to move around in a circle 起旋涡；旋转
be divided into	be grouped into 被分成
be processed into	be made into 被加工成……
as much as possible	to a feasible extent 尽可能地
at the same time	at the same instant 同时
tear into	rip into 撕成
a large number of	massive, extensive, substantial 大量的
so as to	so that 以便
reach a … level	be up to 达到……水平
result in	lead to 导致，结果是
with the help of	with the aid of 在……的帮助下
in parallel	occuring at the same time and having connection 并行的；平行的

◼ Reading Comprehension

Understanding the text

Answer the following questions.

1. What are the basic principles of spinning?

2. What is the yarn made of?

3. Please explain the preliminary processing methods of raw materials.

4. How does the carding function come into being?

5. What is spinning technology classified into?

6. What can the compact spinning be used for?

7. Which spinning has become the most mature and widely used spinning method in the new type of spinning? What are the characteristics?

8. What's the definition of vortex spinning?

9. Why isn't the production technology of compact spinning still very ideal?

10. What are the structural characteristics of vortex spinning yarn?

Critical thinking

Make a Presentation.

During the Qin and Han dynasties, the textile handicraft industry was on a large scale. There was a saying "A man who does not plow will suffer hunger; a woman who does not weave will suffer cold". Work in pairs and try to express what you have known about the textile handicraft industry during the Qin and Han dynasties, then prepare for a presentation in class.

Language Enhancement

Words in use

Fill in the blanks with the words given below. Change the form when necessary. Each word can be used only once.

splice	customize	miscellaneous	preliminary
configuration	adaptability	scour	longitudinal
horizontal	segmentation		

1. He taught me to edit and _____ film.
2. Housework is usually _____ and toilsome.
3. The doctors have successfully concluded _____ tests.
4. We _____ the area for somewhere to pitch our tent.
5. He is currently following more than 600 families from the local community sample in a _____ study.
6. In England, the Saxons are thought to have used both _____ and vertical-axis wheels.
7. Prices range from $119 to $199, depending on the particular _____.
8. Text _____ splits text into individual words or tokens.
9. The _____ of wool is one of its great attractions.
10. You can further _____ these to suit your setup.

Expressions in use

Fill in the blanks with the expressions given below. Change the form when necessary. Each expression can be used only once.

be processed into	with the help of	result in	reach a … level
so as to	tear into	be divided into	in a … sense

1. How can faculty improve their teaching _____ encourage creativity?
2. What might _____ this strange phenomenon?

3. Information gathering has been made easier _____ the Internet.

4. The material will _____ plastic pellets.

5. A typical financial crisis can _____ several stages.

6. Before you _____ that bag of potato chips, drink a glass of water first.

7. _____, the attitude towards strangers that the people have in the city mirrors its warmth.

8. Once they _____ certain _____ of affluence, they also start to buy works of art.

Banked cloze

Fill in the blanks by selecting suitable words from the word bank. You may not use any of the words more than once.

A. rotation	F. unearthed	K. twisting
B. flat	G. mainstream	L. machinery
C. efficiency	H. breakthrough	M. manual
D. achievement	I. pedal	N. stretching
E. spindle	J. pottery	O. embryonic

The most primitive textile tool in China should be spinning brick, which is a flat spinning wheel made of stone or pottery, with a short rod in the middle, making use of the inertia of object 1. _____ to engage in the work of winding and 2. _____ yarn. From the 3. _____ textiles, it can be inferred that there was a spinning wheel in the Spring and Autumn Period. During the Qin and Han dynasties, manual single 4. _____ spinning wheels were widely used. The improvement of the spinning wheel in the Song dynasty moved towards a 5. _____ development, and the emergence of hemp spinning wheel and waterborne spinning wheel actually had the 6. _____ form of small factories. The most important achievement in textile technology is the weaving method and jacquard technology of Yaro brocade forging and other fabrics. The cotton industry is also developing gradually. At the same time, as cotton fabrics jumped to the 7. _____ in the Southern Song dynasty, in order to meet a large number of demand, only the development of faster cotton spinning machines can solve the problem.

A three-spindle 8. _____ spinning wheel appeared in the Yuan dynasty, which could spin three threads at the same time. The three-spinning wheel was a great invention at that time. Before the advent of the machine spinning wheel, it was not easy to find someone who could spin two threads at the same time. The three-spinning wheel not only improved the work 9. _____, but also increased the output. And it is five hundred years earlier than the spinning jenny in Europe. The multi-spindle spinning wheel appeared in the Qing dynasty, which pushed the development of manual textile 10. _____ to a peak. As far as the hand spinning textile industry is concerned, the multi-spindle spinning wheel has become the most complete and fastest spinning wheel to improve efficiency.

■ Translation

I. Translate the following paragraph into Chinese.

Most natural fibers, such as wool, cotton, etc., are only a few inches long, so they must be spun before weaving, and this method was first adopted by the people who lived in the Zagros Mountains of Mesopotamia around 9000 BCE, because they were the first people to tame sheep. At that time, it is probably only necessary to twist with both hands to produce the original woolen yarn. Later, flax and cotton fibers were also used to produce yarn, making the variety of fabrics richer.

II. Translate the following paragraph into English.

　　纺车的出现对纺纱技术具有重大的影响。一般人们都认为，纺车起源于中国，由缫丝机演变而来。纺车在欧洲的起源并不明确，出现的年代较晚。欧洲最早记载纺车的年代约是十四世纪。十五世纪，在撒克逊 (Saxon) 出现了一种经过改进的纺车，燃杆装在机器上，轮子用脚踏板操纵。这种机器可以同时纺纱和绕线，纺纱者的双手得以解放，劳动负担也因此减轻。

Paragraph Writing

How to develop a listing paragraph or essay

Listing is a powerful way to demonstrate a series of observations and to emphasize each element. By listing the writer can analyze causes and effects, state the importance of something, or show the shortcomings or benefits. Listing can also be used to refute opponents' ideas or to state personal opinions.

Most often listing mode starts with general statement and proceeds to details. In other words, the paragraph starts with a topic sentence that states a point of view and then provides discussions usually ordered from the most important idea to the least important, or vice versa.

In Text A of this unit, there are several paragraphs organized in the pattern of listing. Paragraphs of this pattern usually start with a topic sentence followed by pieces of information listed one after another. Have a look at paragraphs 1–5 of Text A, we can find that the author begins with a topic sentence in paragraph 1, then the author lists four types of the basic principles of spinning technology from paragraph 2 to 5.

The basic principles of spinning can be divided into the following four types: removing impurities, releasing, loosing and carding. (para. 1)
The first one is removing impurities…(para. 2)
The second one is releasing…(para. 3)
The third one is loosening…(para. 4)
The fourth one is carding…(para. 5)

There are several points that are worth our attention when we build a paragraph by listing.

First, arrange the supporting ideas in a logical order. It is essential that we should arrange the supporting details in a logical order in everything we write. Otherwise, paragraphs are not effective or cannot be best followed by readers. Chinese students, when writing in English, tend to focus all their attention on expressing ideas and thus ignore the arrangement of the ideas.

Second, we should pay attention to the correct use of listing expressions. The following offers the possible expressions used for listing.

the first point, first, firstly, first of all, to begin with, primarily, initially, other points, second, secondly, next, the next, then, moreover, furthermore, in addition, for the other, another, the other,

last, the last, lastly, in the end, finally, the final, last but not the least...

Remember the following rules when using these expressions.

· Don't mix sets of listing expressions.

· Distinguish "first" from "at first". "First" is a listing word, but "at first" indicates a change.

· Distinguish "last" from "at last". "Last" is a listing word, but "at last" indicates that something happens after a long expectation.

Structured writing

Read the sample paragraph and see how it develops in the pattern of listing.

Topic How to find a good job

Topic sentence

Point 1: Register with more than one agency.

Point 2: Be responsible and trustworthy.

Point 3: Be available all the time.

Sample paragraph

If you want to get the best work possible with an employment agency, it is important to remember the following three points. First, register with more than one agency so that you have a better chance of getting a job. Second, be responsible and trustworthy. When you receive an assignment, do your best. Employment agencies usually keep a record of your employment history, which may affect your future job-seeking. Finally, because employment agencies may contact you at any time, you need to be available all the time. Making yourself available and ready is very important to get the suitable work.

Write a paragraph on one of the following topics in the pattern of listing. One topic has an outline that you can follow.

Topic

Why people get fired from their jobs

Topic sentence

There are several reasons why people get fired from their jobs.

Reasons

dishonesty: cheating in job application, telling lies in work, etc.

poor attendance: late for work or frequently absent

bad teamwork spirit: having difficulty getting along with other people

More topics:

Topic 1: Benefits of travelling

Topic 2: Advantages of shopping online

Text B

Knitted Fabrics

1   **Knitting** is a fabric formed by bending yarn into **loops** and stringing each other with knitting needles. The difference between knitted fabrics and woven fabrics lies in the shape of the yarn in the fabric. Knitting is divided into weft knitting (weft knitted fabric) and warp knitting (warp knitted fabric).

2   Weft knitted fabrics are often made of low **elastic** polyester **filament** or special-shaped polyester filament, nylon filament, cotton yarn, wool yarn, etc., using flat needle organization, variable flat needle organization, **rib** weave, double rib weave, **jacquard** weave, **terry** weave and so on. There are many kinds of weft knitted fabrics, which generally have good

elasticity and **extensibility**. The fabric is soft, wrinkle-resistant and easy to wash and dry. The main disadvantages are that the fabric has poor **hygroscopicity**, is not stiff enough, and easy to **break off** and crimp. Chemical fiber fabric tends to fuzz, pill and hook wire. Weft knitted fabrics have the following types: polyester yarn-dyed knitted fabric, polyester knitted labor fabric, polyester knitted core fabric, polyester cotton knitted fabric, artificial fur knitted fabric, velvet knitted fabric and **port** knitted wool.

3   Warp knitted fabrics are often made of polyester, nylon, **polypropylene** and other synthetic filaments, as well as cotton, wool, silk, hemp, chemical fiber and their blended yarn. Ordinary warp knitted fabrics are often woven with chain weave, warp flat weave, warp **satin** weave, warp **oblique** weave and so on. There are many kinds of fancy warp knitted fabrics, such as **mesh** fabric, terry fabric, **pleated** fabric, **plush** fabric, weft fabric and the like. Warp knitted fabric has the advantages of good longitudinal dimensional stability, **stiff** fabric, small **dispersion**, no crimping, good air **permeability**, etc. **Lateral** extension, elasticity and softness are not as good as weft knitted fabrics. Warp knitted fabrics have the following types: polyester warp knitted fabric, warp knitted pile fabric, warp knitted mesh fabric, warp knitted velvet fabric, warp knitted loop fabric and warp knitted jacquard fabric.

4   What is the difference between warp knitting and weft knitting? Warp knitting uses multiple threads to form loops along the longitudinal (warp direction) of the cloth at the same time. Weft knitting uses one or more threads to be looped along the transverse (weft) of the department. Weft knitted fabrics can be formed with at least one thread, and multiple threads are generally used to weave in order to improve production efficiency, whereas warp knitted fabrics cannot be formed with one thread. A thread can only form a curl made up of one **coil**. All weft knitted fabrics can **be separated into** threads against the knitting direction, but warp knitted fabrics cannot. Warp knitted fabrics cannot be knitted by hand.

5   There are some points for attention in clothing design when using knitted fabrics.

6   **To begin with**, we should make use of the flexibility of knitted fabrics. Knitted fabric has good flexibility, which can minimize the **seams**, pleats, splicing and so on. In addition, knitted fabrics **are** generally not **suitable for** the use of push-back, pull-hot skills modeling, but the use of the elasticity of the fabric itself or the appropriate use of wrinkling techniques to suit the curve of the human body. Then the size of the fabric flexibility has become an important basis in the **template** design and production. In the end, the sample of woven clothing is generally larger than the area needed to wrap the human body, that is, there is a certain amount of loosening relative to the human body, while knitted clothing is different from the structure of the fabric used. If the fabric with great elasticity (related to the yarn and tissue structure) is designed without loosening, its model size can not only be the same as the human **circumference** size, but also can be reduced by considering the **coefficient** of elasticity.

7   Second, we should make use of the crimping property of knitted fabrics. The crimping property of knitted fabric is the winding phenomenon of edge fabric caused by the disappearance of internal stress in the edge coil of knitted fabric. Crimping is the **deficiency** of knitted fabrics. It can cause **uneven** seams or changes in the size of the edges of the clothing, and **ultimately** affect the overall modeling effect of the clothing and the size of the clothing. However, not all knitted fabrics have crimping, but only fabrics with individual structures, such as weft flat knitted fabrics, can be solved by adding dimensions for edge drawing, ribbing or rolling and inlaying adhesive interlining at the edge of the garment. The crimping of some knitted fabrics has been **eliminated** in the process of fabric finishing, which avoids the trouble of template design. It should **be pointed out** that **on the basis of** understanding the performance of the fabric, many designers can make use of the crimping property of the fabric to design at the **neckline** and **cuff** of the model, so that the clothing can get a special

appearance style, which is refreshing, especially in the knitting of molded clothing, and its crimping property can also be used to form a unique pattern or split line.

8   Third, the **dispersivity** of knitted fabrics should be paid attention to. Knitted fabrics are different from woven fabrics in style and characteristics. The style of knitted fabrics should not only emphasize **giving full play to** the advantages of fabrics, but also **overcome** their disadvantages. As individual knitted fabrics are **discrete**, too much **exaggeration**, darts design, cutting lines, and seams shouldn't be used in the design and production of some knitted fabrics. In order to prevent the **disintegration** of knitting coils and affect the clothing **serviceability**, simple and soft lines should be used to **coordinate** with the soft fit style of knitted fabrics.

Notes

low elastic polyester filament   低弹涤纶丝

special-shaped polyester filament   异型涤纶丝

flat needle organization   平针组织

variable flat needle organization   变化平针组织

rib weave   罗纹组织

double rib weave   双罗纹组织

jacquard weave   提花组织

terry weave   毛圈组织

inlaying adhesive interlining   镶嵌粘合衬

New words and phrases

knit /nɪt/

v. to make clothes, etc. from wool or cotton thread using two long thin knitting needles or a machine 编织；针织；机织

loop /luːp/   *n.* a shape like a curve or circle made by a line curving right round and crossing itself 环形；环状物；圆圈

v. to form or bend sth into a round 使成环；使绕成圈

elastic /ɪˈlæstɪk/   *n.* material made with rubber, that can stretch and then return to its original size 橡皮圈（或带）；松紧带

a. able to stretch and return to its original size and shape 有弹性的；有弹力的

filament /ˈfɪləmənt/   *n.* a thin wire in a light bulb that produces light when electricity is passed through it （电灯泡的）灯丝；丝极

rib /rɪb/   *n.* any of the curved bones that are connected to the spine and surround the chest 肋骨

jacquard /ˈdʒækɑːd; -kəd/   *n.* a fabric in which the design is incorporated into the weave instead of being printed or dyed on 提花织物

terry /ˈteri/   *n.* a type of soft cotton cloth that absorbs liquids and has a surface covered with raised loops of thread, used especially for making towels 毛圈棉织物（多用来做毛巾）

extensibility /ɪkˌstensəˈbɪlətɪ/   *n.* the ability to be extended or stretched 展开性；可延长性

hygroscopicity /ˌhaɪɡrəskəˈpɪsɪtɪ/   *n.* the stance of readily taking up and retaining moisture 吸水性；吸湿性

port /pɔːt/   *n.* a town or city with a harbor, especially one where ships load and unload goods 港口城市；港市；口岸

polypropylene /ˌpɒlɪˈprəʊpəliːn/   *n.* a strong plastic often used for objects such as toys or chairs that are made in a mould 聚丙烯

satin /ˈsætɪn/   *n.* a type of cloth with a smooth shiny surface 缎子

a. having the smooth shiny appearance of satin 缎子似的；平滑而有光泽的

oblique /əˈbliːk/   *a.* ① not expressed or done in a direct way 间接的；不直截了当的；拐弯抹角的 ② sloping at an angle 斜的；倾斜的

mesh /meʃ/   *n.* material made of threads of plastic rope or wire that are twisted together like a net 网状物；网状织物

v. to fit together or match closely, especially in a way that works well; to make things fit together successfully（使）吻合；相配；匹配，适合

pleated /ˈpliːtɪd/   *a.* having pleats 有褶的

plush /plʌʃ/   *n.* a type of silk or cotton cloth with a thick soft surface made of a mass of threads（丝或棉的）长毛绒

a. very comfortable; expensive and of good quality 舒适的；豪华的

stiff /stɪf/   *a.* firm or not to bend easily 硬的；不易弯曲的

dispersion /dɪˈspɜːʃn/   *n.* the process by which people or things are spread over a wide area 分散；散开；散布

permeability /ˌpɜːmɪəˈbɪləti/   *n.* the state or quality of being permeable 渗透性；可透性

lateral /ˈlætərəl/   *a.* connected with the side of sth or with movement to the side 侧面的；横向的

coil /kɔɪl/   *n.* a series of circles formed by winding up a length of rope, wire, etc.（绳子或金属线的）圈

v. to wind or gather (ropes, hair, etc.) into loops or (of rope, hair, etc.) to be formed in such loops 将（绳、头发等）盘成圈

seams /siːmz/   *n.* a line of stitches which joins two pieces of cloth together 缝合线

template /ˈtempleɪt/   *n.* a thing that is used as a model for producing other similar examples 样板；模框；标准

circumference /səˈkʌmfərəns/   *n.* a line that goes around a circle or any other curved shape 圆周；圆周长

coefficient /ˌkəʊɪˈfɪʃnt/   *n.* a number that measures a particular property of a substance （测定物质某种特性的）系数

a. several people or groups of people working together 合作的；协作的；协力的

deficiency /dɪˈfɪʃnsi/   *n.* the state of not having, or not having enough of sth that is essential 缺乏；缺少；不足

uneven /ʌnˈiːvn/   *a.* not level, smooth or flat 凹凸不平的；不平坦的

ultimately /ˈʌltɪmətli/   *ad.* in the end; finally 最终；最后；终归

eliminate /ɪˈlɪmɪneɪt/   *v.* ① to remove or get rid of sth/sb 排除；清除；消除 ② [usually passive] to defeat a person or a team so that they no longer take part in a competition, etc. （比赛中）淘汰

neckline /ˈneklaɪn/   *n.* the edge of a piece of clothing, especially a woman's, which fits around or below the neck （女装的）领口；开领

cuff /kʌf/   *n.* the end of a coat or shirt sleeve at the wrist 袖口

v. to hit sb quickly and lightly with your hand, especially in a way that is not serious 用手轻快地拍（某人）

dispersivity /dɪspəˈsɪvɪtɪ/   *n.* the state or the degree of dispersion 分散性

overcome /ˌəʊvəˈkʌm/	*v.* to succeed in dealing with or controlling a problem that has been preventing you from achieving sth 克服；解决
discrete /dɪˈskriːt/	*a.* independent of other things of the same type 分离的；互不相连的
exaggeration /ɪɡˌzædʒəˈreɪʃn/	*n.* a statement or description that makes sth seem larger, better, worse or more important than it really is; the act of making a statement like this 夸张；夸大；言过其实
disintegration /dɪsˌɪntɪˈɡreɪʃn/	*n.* split 分裂，分解；瓦解，解体
serviceability /ˌsɜːvɪsəˈbɪlɪtɪ/	*n.* the quality of being able to provide good service 可用性，适用性；使用可靠性
coordinate /kəʊˈɔːdɪneɪt/	*v.* to organize the different parts of an activity and the people involved in it so that it works well 使协调；使相配合
break off	become severed 折断
be separated into	be divided into 被分离成
to begin with	at first 首先，起初
be suitable for	fit for 适合于……的
be pointed out	be made or written a comment 被指出
on the basis of	be based on/upon 根据；基于……
give full play to	fully functioning 充分发挥

▪ Reading Comprehension

Understanding the text

Answer the following questions.

1. What are the two types of knitted fabrics?
   A. Weft knitting or weft knitted fabric.

B. Warp knitting or warp knitted fabric.

C. Both A and B.

D. None of above.

2. What are weft knitted fabrics often made of?

A. Low elastic polyester filament.

B. Special-shaped polyester filament.

C. Nylon filament, cotton yarn, wool yarn, etc.

D. All of above.

3. The advantages of warp knitted fabric includes _____.

A. low elastic polyester filament

B. good longitudinal dimensional stability and stiff fabric

C. small dispersion, no crimping, good air permeability

D. both B and C

4. Velvet knitted fabric belongs to _____.

A. low elastic polyester filament

B. stiff fabric

C. weft knitted fabrics

D. warp knitted fabrics

5. Which of the following is correct?

A. Weft knitted fabrics can be formed with at least one thread, but in order to improve production efficiency, multiple threads are generally used to weave.

B. Warp knitted fabrics can be formed with one thread.

C. All weft knitted fabrics can't be separated into threads against the knitting.

D. Warp knitted fabrics can be knitted by hand.

Critical thinking

Work in pairs and discuss the following questions.

1. Please explain port knitted wool and its characteristic.

2. What should we pay attention to in clothing design when using knitted fabrics?

3. How many types do weft knitted fabrics have?

Language Enhancement

Words in use

Fill in the blanks with the words given below. Change the form when necessary. Each word can be used only once.

coordinate	exaggeration	eliminate	deficiency
overcome	oblique	ultimately	disintegration
neckline	extensibility		

1. Vitamin _____ in the diet can cause illness.
2. _____, you'll have to make the decision yourself.
3. Recent measures have not _____ discrimination in employment.
4. She wear a dress with a low _____.
5. She _____ injury to win the Olympic gold medal.
6. It would be an _____ to say I knew her well—I only met her twice.
7. The report describes the catastrophic _____ of the aircraft after the explosion.
8. Government officials visited the earthquake zone on Thursday morning to _____ the relief effort.
9. Figure 2 shows the flexibility and _____ provided by using security plug-ins.
10. This is an _____ reply to a simple question.

Expressions in use

Fill in the blanks with the expressions given below. Change the form when necessary. Each expression can be used only once.

give full play to	be pointed out	a variety of	be suitable for
on the basis of	be separated into	to begin with	break off

1. He resigned for _____ reasons.
2. Britain threatened to _____ diplomatic relations.

3. Physiographically, the United States may _____ three major divisions.

4. It must _____ that the statistics are inaccurate.

5. I need a job that can _____ my skills.

6. Statement should be made _____ fact.

7. These regions seem too dry to _____ farming.

8. _____, the first step for the interview today is to prepare yourself with your identity card and admission card.

Sentence structure

Complete the following sentences by translating the Chinese into English, using "whereas ..." structure.

Model:	Weft knitted fabrics can be formed with at least one thread, _____ (然而经纱编织物用一根纱线是无法形成的).
	→Weft knitted fabrics can be formed with at least one thread, whereas warp knitted fabrics cannot be formed with one thread.

1. Generally speaking, the virus is not easy to mutate, _____ _____ (然而这种病毒就很容易变异).

2. The vast majority of college students can get a diploma after four years of study, _____ (不过极个别学生因考试不及格拿不到毕业证).

3. Lawrance's very good at science, _____ _____ (而他的兄弟绝对是不可救药).

Rewrite the following sentences by using "be not as good as ..."

Model:	I hoped my scores would be good, but they were not.
	→My scores were not as good as I had hoped.

1. His Arabic is not so good compared with his English.

_____.

2. The first movie in a series is often good, but the second one is not so good.

_____.

3. His appetite was good, but it had been better.

_____ .

Air-Jet Spinning and Air-Jet Weaving Technology

1   Air-jet spinning is a new spinning method which uses high-speed rotating air flow to twist the yarn into yarn. Air-jet spinning adopts sliver feeding, four-roller double short apron super-large draft, twisted into yarn by fixed nozzle. After the yarn is drawn out, it is wound to the bobbin through the yarn cleaner and wound directly into bobbin yarn. Air-jet spinning can spin 30-7.4tex yarns, which is suitable for pure spinning and blending of chemical fiber and cotton. Because of the special yarn forming mechanism of air-jet spinning, the structure and properties of air-jet yarn are obviously different from those of ring-spun yarn, and its products have a unique style.

2   There are such characteristics of air-jet spinning as high spinning speed, short technological process, good quality and distinctive product, as well as adapting to a wide range of varieties.

3   Air-jet spinning uses air twisting and no high-speed rotary parts (such as steel rings in ring spinning) to achieve high-speed spinning. The spinning speed can reach 120–300 m/min, and the output per head is 10–15 times that of ring spinning. Compared with ring spinning, air-jet spinning has less roving and bobbin processes, and saves about 30% of the plant area.

4   Compared with ring spinning, ten thousand spindles employ 90 people, which is reduced by about 60%, and the machine material consumption is about 30% lower than that of ring spinning, and the usual maintenance cost and maintenance workload are also reduced.

5　　The comprehensive evaluation of the quality of air-jet yarn is good, except that the yarn strength is 5% lower than that of ring spinning and other quality indexes are better than ring spinning. The physical properties of air-jet yarn, such as CV value of evenness, coarse details and yarn defects, are better than those of ring-spun yarn. The hairiness above 3mm is less than that of ring-spun yarn. Although the yarn strength is lower, the strength unevenness of ring-spun yarn is lower than that of ring-spun yarn. It is suitable for the weaving of new looms such as rapier looms and air-jet looms, and the production efficiency of the looms can be increased by more than 2%. The quality of air-jet spinning is not only similar to that of ring spinning, but also has its uniqueness. The friction coefficient of air-jet yarn is large. The yarn is directional, and its friction performance is also directional. The wear resistance of ring-spun yarn is better than that of ring-spun yarn, but the handle is harder. With the addition of some devices in air-jet spinning, some special kinds of yarns can also be developed, such as fancy yarn, core-spun yarn, mixed yarn and so on.

6　　Air-jet yarn can be used not only for knitting products, but also for woven products. More applications are knitted T-shirt products which are no skew, less shadow, less strip defects, bedding, leisure products, such as two-sided casual wear, sportswear and so on. According to the characteristics of air-jet yarn, we can also develop products with unique style. As for the use of air-jet fabric stiff, rough and other characteristics, it will be processed into linen-like fabric, as well as crepe fabric, wool-like products and so on.

7　　Air-jet loom has become one of the most promising new looms in the global weaving market because of its reasonable air-jet weft insertion mode, simple and safe operation and high weft insertion rate. In recent years, the world's well-known manufacturers have invested in research and development in expanding the weaving range of air-jet looms, improving fabric quality, reducing energy consumption and increasing speed, so that it can weave untwisted yarn, strong twist yarn, elastic yarn, web yarn, fancy yarn, different yarn fabric and so on.

8   The main representatives of foreign air-jet looms include: (Picanol) Omni and DELTA-190; Toyota JA-T610 series and JA2S-280; Tsuda Kui (Tsudakoma) ZAX9100, ZA203-280 and ZAX-360 and so on. Among them, Japanese air-jet looms are mainly negative cam openings, while European air-jet looms are mainly electronic Dobby openings.

9   Protec (Promatech), Biga Le, Toyota and other products highlight the integrity and damping effect of mechanical structure design, improve the integrity and rigidity of the frame by improving the structure of the basic components and connectors of the frame, and adopt short-range beating-up mechanism and lightweight design to improve the stability of the loom, so the loom can be directly placed on the shock cushion, and the feet do not need bolts or bonding.

10   The nozzles of air-jet looms have been developed to 2 or 4, or more nozzles, and the production speed is further improved. The jet action of the nozzle is completed by the opening and closing movement of the solenoid valve controlled by the computer, and the air pressure pulse is stable, so the ultra-high speed stable weft insertion is realized. The pressure of the main nozzle can be set according to the yarn number, and the pressure of the auxiliary nozzle can be switched according to the high and low stages to produce a fabric with a weft yarn number difference of up to 12 times.

11   Micro-computer technology is widely used in new air-jet looms. 32-bit CPU and 16-bit CPU groups form a network through optical fiber, which reduces or shortens the mechanical chain and improves the degree of mechatronics of the loom and the accuracy of various movements. Some looms also reserve an Ethernet interface to emphasize the remote production organization and management of the factory. Equipped with weft tension controller, the weft insertion speed can be reduced at the end of weft insertion, so as to reduce the peak weft tension and reduce weft breakage, thus weaving extra-wide fabrics, elastic fabrics, wool fabrics and lower strength weft yarns. The main transmission parts are in the form of oil bath lubrication, and most of the other refueling parts use centralized refueling with controllable oil quantity,

so as to ensure that the whole machine is in good lubrication condition when the loom is running at high speed.

12   Generally speaking, European air-jet looms have abandoned the original wallboard structure and adopted a solid box structure, and heavy-duty transmission parts such as beating-up conjugate cams are assembled in the box body to enhance the overall strength and weight of the frame, and the vibration of the loom is reduced to a minimum. heavy fabrics with asymmetric structure and multi-page Heald frames can be processed.

13   As for domestic air-jet looms, the main representative models are Xianyang loom gax and ZA209i, medium spinning machine GA718, Suzhou spinning machine ASGA701 and so on. Among them, Xianyang loom ZA209i air-jet loom is produced in cooperation with Tsuda Kui. It is a series of machines with 210cm, 230cm, 280cm and 330cm sley width, equipped with multi-arm and active opening device, and its running efficiency is close to the performance standard of Tsuda Kui ZA209i air-jet loom. The main changes are: the motor adopts 3.0kW external type; the opening adopts ZCM-III cam box, which increases the strength of the opening mechanism parts and improves the adaptability of the opening system to high speed; the beating-up rocking shaft increases the middle support, improves the stiffness of the beating-up system, and the beating-up movement becomes more stable; the main nozzle adopts the combination of solenoid valve and L-type main nozzle, which improves the response speed of the main injection system. The wallboard structure, the front and rear bracing and the breast beam strength are improved, and the brake torque is increased to 588N · m, which can realize high-speed production.

Unit 3

Fashion Design

The best color in the whole world is the one that looks good on you .
—*Coco Chanel*

Fashion is part of our culture, and it's about more than just a pretty dress.
—*Joan Smalls*

Pre-Reading Activities

1. Listen to the recording and answer the following questions.

(1) What is fashion?

(2) What is a fashion designer?

2. Listen to the recording again and fill in the blanks based on what you hear.

Fashion is something we deal with everyday. Fashion is a means of self-expression that allows people to try on many roles in life. Whether you prefer hip-hop or Chanel-chic, fashion (1) _____ the chameleon in all of us. It's a way of celebrating the diversity and variety of the world in which we live. Fashion is about change which is necessary to keep life interesting. It's also a mirror of sorts on society. It's a way of measuring a mood that can be useful in many aspects, culturally, socially even (2) _____.

A fashion designer creates clothing, including dresses, suits, pants, and skirts, and (3) _____ like shoes and handbags, for consumers. He or she can (4) _____ in clothing, accessory, or jewelry design, or may work in more than one of these areas.

According to the U.S. (5) _____ of Labor Statistics, the growth outlook for fashion designers seeking to work with (6) _____ manufacturers over the next (7) _____ relative to other occupations and industries is slower than the average for all (8) _____.

3. Discuss the following questions with your partner.

(1) Who dictates fashion? Namely, musicians and other cultural icons? How do they influence people's wearing?

(2) How to be a fashion designer?

The Concept and Elements of Fashion Design

1   What is design? The so-called design is the act of expressing intention based on **prior** consideration from the selection of material through the entire production process to the completion and use of the work. The term "design" covers a wide range, which can be divided into three major categories: product design including textiles design, vehicle design and daily necessities design; space design, including urban design and landscape design; visual design, including graphic design, display design, film and television design, and **animation** design. The fashion design belongs to the first category, namely the product design. Design is an important link in the process of **costume** art creation. Fashion design is an artistic behavior based upon the objective demand of a certain material and social concept in a variety of materials and their restrictive relations to be selected, refined, **decomposed** and **synthesized** through human mind.

2   The characteristics of comprehensive discipline and edge discipline of design are **prominent** in fashion design: it is first carried out on the human body, which involves many fields such as **ergonomics, physiology, anatomy, anthropology**. Then, people in clothing live in space and environment, which covers the fields of environmental **ecology**, geography, **meteorology**, biology, **cosmic** science. Human beings are advanced creatures with thinking ability, hence fashion design is affected by such disciplines as physiology, **ethics**, religion, art theory, history, politics, social psychology, and economics. Overall, the design and implementation of fashion is a complex process. which has both artistic and practical attributes.

3　　Fashion is both an art and a science. It is very personalized and **incredibly** popular. These seemingly **antagonistic** opposing factors of fashion are **dialectically** unified and **mutually** attracted and **simulated**, thus arousing the public desire to buy clothes. Meanwhile, it is the combination of all these contradictory elements that makes the fashion business keep ever-changing and colorful vitality. The reason why fashion can **be viewed as** an art is that so much creativity, the source of fashion design, is required in designing. Unlike most other businesses where **conformity** is the norm, innovation and creation are the vitality of development in clothing industry. Fashion has also been regarded as a science since clothing could never be mass-produced without machines and equipment. Technology has **contributed to** the development of clothing production methods. It can be said that almost all stages from fashion design to sales **are** more or less **dependent on** science and technology, especially in the era of intelligent manufacturing.

4　　Clothing design works are never created randomly. An excellent fashion design is an **ingenious** application of various factors to a clothing. **Silhouette**, detail, texture and color are the four basic elements that **are indispensable to** fashion design. Only through a change in one or more of these basic components does a new fashion evolve.

5　　Fashion silhouette, the backlight silhouette effect of fashion, refers to the overall outline or **couture**, also frequently known as "**profile**" or "shape or form". It is the first visual element of fashion style, and the first factor to be considered in fashion design. Fashion silhouette can **be summarized into** three basic forms: straight or **tubular**, bell-shaped or **bouffant**, and **bustle** or back fullness—with many variations for a particular design style.

6　　In the colorful fashion, you will be surprised to find by careful observation that various fashions are charming and attractive just because of their different details such as collar, sleeve, pocket, **front fly** and waistline. The personalized factors that give the silhouette its form or shape are called details, which include tailoring of collars, sleeves and shoulders, changes in length and width of skirts or trousers/pants, and trimmings. By changing the details, the silhouettes of fashion evolve gradually from one style to another. Variations in the details allow both designers and consumers to express their own individuality freely. To highlight a natural waistline silhouette, for instance, a slender lady might choose a simple wide belt, a decorative belt, or a belt in a contrasting color with the dress.

7   The selection and application of fashion material is a significant part for fashion design. It is very important for fashion designers to choose and use material properly. Different fashion styles need to be represented by different fabrics. There are thousands of different styles and thousands of materials. Different styles can be obtained by using different modeling and technique. The intrinsic factor closely related to the fashion material is texture, which can affect the appearance of a silhouette, giving it a bulky or slender look, depending on the roughness or smoothness of the materials. Texture directly affects the drape of a dress. Chiffon is close-fitting and flowing, making it a good choice for soft, feminine styles, while corduroy is suitable for more casual wear for its firmness and bulk.

8   It is well known that fashion design cannot be separated from color. As the saying goes, one can see the color afar and the flowers near on clothes. The most important element for textiles is color, which is added to enhance the fabric's appeal. Color symbolism often changes with geographical location. In the western nations, white is the symbol of purity, worn by brides and used in communion dresses; but in India, it is the color of mourning. Historically, color had been used to denote rank and occupation. For example, purple had always been associated with royalty, and in aristocratic dress during certain periods. Color has always been a major consideration in women's fashion design. Color is a key factor to be considered when people select clothes. A designer's color palette varies with different seasons and consumer's preferences.

Notes

front fly　裤门襟；门襟；前门襟

New words and phrases

prior /ˈpraɪə(r)/

a. happening or existing before sth else or before a particular time 先前的；较早的；在前的

◇ **prior to** (*formal*) before sth 在……之前

animation /ˌænɪˈmeɪʃn/

n. ① [U] energy and enthusiasm in the way you look, behave or speak 生气；活力；富有生命力　② [U] the process of making films/movies, videos and computer games in which drawings or models of people and animals seem to move（指电影、录像、电脑游戏的）动画制作 ③ [C] a film/movie in which drawings of people and animals seem to move 动画片

costume /ˈkɒstjuːm/

n. ① [C, U] the clothes worn by people from a particular place or during a particular historical period（某地或某历史时期的）服装；装束 ② [C, U] the clothes worn by actors in a play or film/movie, or worn by sb to make them look like sth else（戏剧或电影的）戏装；服装

decompose /ˌdiːkəmˈpəʊz/

v. ① to be destroyed gradually by natural chemical processes 腐烂② ~ **(sth) (into sth)** to divide sth into smaller parts; to divide into smaller parts（使）分解

synthesize /ˈsɪnθəsaɪz/

v. ① (*technical* 术语) to produce a substance by means of chemical or biological processes（通过化学手段或生物过程）合成② to produce sounds, music or speech using electronic equipment（音响）合成③ to combine separate ideas, beliefs, styles, etc. 综合

prominent /ˈprɒmɪnənt/ *a.* ① important or well known 重要的；著名的；杰出的 ② easily seen 显眼的；显著的；突出的

ergonomics /ˌɜːɡəˈnɒmɪks/ *n.* [U] the study of how equipment and furniture can be arranged so that people can do work or other activities more efficiently and comfortably 工效学

physiology /ˌfɪziˈɒlədʒi/ *n.* [U] the scientific study of how people's and animals' bodies function, and of how plants function 生理学

anatomy /əˈnætəmi/ *n.* [U] the study of the structure of the bodies of people or animals 解剖学

anthropology /ˌænθrəˈpɒlədʒi/ *n.* [U] the study of the human race, especially of its origins, development, customs and beliefs 人类学

ecology /iˈkɒlədʒi/ *n.* [U] the relation of plants and living creatures to each other and to their environment; the study of this 生态；生态学

meteorology /ˌmiːtiəˈrɒlədʒi/ *n.* [U] the scientific study of the earth's atmosphere and its changes, used especially in forecasting the weather 气象学

cosmic /ˈkɒzmɪk/ *a.* [usually before noun] ① connected with the whole universe 宇宙的 ② occurring in, or coming from, the part of space that lies outside Earth and its atmosphere 外层空间的 ③ very great and important 巨大且重要的

ethics /ˈeθɪks/ *n.* [U] the study of questions about what is morally right and wrong 伦理学

incredibly /ɪnˈkredəbli/ *ad.* ① extremely 极端地；极其 ② in a way that is very difficult to believe 令人难以置信

antagonistic /ænˌtæɡəˈnɪstɪk/ *a.* ~ **(to/toward(s) sb/sth)** *(formal)* showing or feeling opposition 对立情绪的；对抗的；敌对的；敌意的

dialectically /ˌdaɪəˈlektɪkli/   *ad.* 辩证法地

mutually /ˈmjuːtʃuəli/   *ad.* felt or done equally by two or more people 相互地；彼此；共同地

simulate /ˈsɪmjuleɪt/   *v.* ① to be made to look like sth else 模仿；冒充 ② to create particular conditions that exist in real life using computers, models, etc., usually for study or training purposes（用计算机或模型等）模拟 ③ to pretend that you have a particular feeling 假装；冒充；装作

contradiction /ˌkɒntrəˈdɪkʃn/   *n.* [C, U] ~ (between A and B) a lack of agreement between facts, opinions, actions, etc.（事实、看法、行动等的）不一致；矛盾；对立

conformity /kənˈfɔːməti/   *n.* [U] ~ (to/with sth) (*formal*) behavior or actions that follow the accepted rules of society（对社会规则的）遵从；遵守

intelligent /ɪnˈtelɪdʒənt/   *a.* (*computing* 计) (of a computer, program, etc. 计算机、程序等) able to store information and use it in new situations 智能的

random /ˈrændəm/   *a.* [usually before noun] done, chosen, etc. without sb deciding in advance what is going to happen, or without any regular pattern 随机的；随意的

n. (**at random**)without deciding in advance what is going to happen, or without any regular pattern 随意；随机；胡乱

ingenious /ɪnˈdʒiːniəs/   *a.* ① (of an object, a plan, an idea, etc. 物体、计划、思想等) very suitable for a particular purpose and resulting from clever new ideas 精巧的；新颖独特的；巧妙的 ② (of a person 人) having a lot of clever new ideas and good at inventing things 心灵手巧的；机敏的；善于创造发明的

silhouette /ˌsɪluˈet/ *n.* ① [C, U] the dark outline or shape of a person or an object that you see against a light background（浅色背景衬托出的）暗色轮廓 ② [C] the shape of a person's body or of an object（人的）体形；（事物的）形状

couture /kuˈtjʊə(r)/ *n.* [U] *(from French)* the design and production of expensive and fashionable clothes; these clothes 时装设计制作；高级时装

profile /ˈprəʊfaɪl/ *n.* ① the outline of a person's face when you look from the side, not the front 面部的侧影；侧面轮廓 ② a description of sb/sth that gives useful information 概述；简介；传略

tubular /ˈtjuːbjələ(r)/ *a.* made of tubes or of parts that are shaped like tubes 管子构成的；管状的

bouffant /buːˈfɑːnt/ *a.* (of a person's hair 头发) in a style that raises it up and back from the head in a high round shape（往后梳）蓬松式的

bustle /ˈbʌsl/ *n.* ① [C] a frame that was worn under a skirt by women in the past in order to hold the skirt out at the back（旧时女子用的）裙撑 ② [U] busy and noisy activity 忙乱嘈杂；喧闹

modeling /ˈmɒdəlɪŋ/ *n.* the act or an instance of making a model 模型制造

intrinsic /ɪnˈtrɪnzɪk; ɪnˈtrɪnsɪk/ *a.* ~ **(to sth)** belonging to or part of the real nature of sth/sb 固有的；内在的；本身的

texture /ˈtekstʃə(r)/ *n.* the way a surface, substance or piece of cloth feels when you touch it, for example how rough, smooth, hard or soft it is 质地；手感

bulky /ˈbʌlki/ *a.* ① (of a thing 东西) large and difficult to move or carry 庞大的；笨重的 ② (of a person 人) tall and heavy 大块头的；高大肥胖的

drape /dreɪp/	*n.* a long thick curtain（厚且长的）帘子；帷帘；帷幕
chiffon /ˈʃɪfɒn/	*n.* [U] a type of fine transparent cloth made from silk or nylon, used especially for making clothes 雪纺绸；薄绸；尼龙绸（尤用于制衣）
close-fitting /ˌkloʊs ˈfɪtɪŋ/	*a.* (of clothes 衣服) fitting tightly, showing the shape of the body 紧身的
feminine /ˈfemənɪn/	*a.* having the qualities or appearance considered to be typical of women; connected with women（指气质或外貌）女性特有的；女性的；妇女的
corduroy /ˈkɔːdərɔɪ/	*n.* (also cord) [U] a type of strong soft cotton cloth with a pattern of raised parallel lines on it, used for making clothes 灯芯绒
afar /əˈfɑː(r)/	*ad.* a long distance away 远处地；遥远地
communion /kəˈmjuːniən/	*n.* [C] (*technical* 术语) a group of people with the same religious beliefs 教派；教会；宗教团体
mourning /ˈmɔːnɪŋ/	*n.* ① sadness that you show and feel because sb has died 伤逝；哀悼 ② clothes that people wear to show their sadness at sb's death 丧服
denote /dɪˈnəʊt/	*v.* ① to be a sign of sth 标志；预示；象征 ② to mean sth; represent 表示；意指
palette /ˈpælət/	*n.* [usually sing.] (*technical* 术语) the colors used by a particular artist（画家使用的）主要色彩；主色调
aristocratic /ˌærɪstəˈkrætɪk/	*a.* belonging to or typical of the aristocracy 贵族的
be viewed as	to think about sb/sth in a particular way 把……视为；以……看待
contribute to	to be one of the causes of sth 是……的原因之一

be dependent on	determined by conditions or circumstances that follow 依赖于；依靠
be indispensable to	too important to be without to sb/sth 对……必不可少的
be summarized into	to give a summary of sth 归纳为；归结为
be associated with	to make a connection between people or things in your mind 与……有联系 / 关联
be added to	to put sth together with sth else so as to increase the size, number, amount, etc. 增加；添加
evolve from	to develop gradually, especially from a simple to a more complicated form; to develop sth in this way （使）逐渐形成；逐步发展；逐渐演变
vary with	to change or be different according to the situation （根据情况）变化；变更；改变

■ Reading Comprehension

Understanding the text

Answer the following questions.

1. What is design? What is fashion design?

2. The characteristics of comprehensive discipline and edge discipline of design are prominent in fashion design. Give some examples to illustrate it.

3. Why is fashion both an art and a science?

4. What basic elements does fashion design cover?

5. Based on the passage, list the three basic forms of fashion silhouette.

6. When a detail is designed to reach an extreme, a reversal trend takes place. Why?

7. What is one of the most significant features of fashion texture?

8. Why does a designer's color palette vary with different seasons and consumer's preferences?

Critical thinking

I. Make a Presentation.

Based on the introduction to fashion design, find articles or stories relevant to the theme of a fashion designer. Pick up one or two inspiring stories about some famous fashion designers and make a presentation in class.

II. Work in pairs and discuss the following questions.

1. What are required to be a fashion designer?

2. What is a fashion designer's job like in your mind?

3. Do you want to be a designer after graduation? Why or why not?

■ Language Enhancement

Words in use

Fill in the blanks with the words given below. Change the form when necessary. Each word can be used only once.

contradiction	ingenious	enhance	conformity	simulate
synthesize	feminine	prominent	intrinsic	denote

1. After extensive research, Albert Hoffman first succeeded in _____ the acid in 1938.

2. Friendships are not made in a day, and the computer would be more acceptable as a friend if it _____ the gradual changes that occur when one person is getting to know another.

3. Diamonds have little _____ value and their price depends almost entirely on their scarcity.

4. Internet advertising will play a more _____ role in organizations' advertising in the near future.

5. The figures that were released recently _____ that we have for the first time in more than a year seen positive economic growth.

6. Some people prefer not to use the _____ form "actress" and use the word

"actor" for both sexes.

7. Researchers cannot reach a conclusion because there is a _____ between the two sets of figures.

8. The prime minister is, in _____ with their constitution, chosen by the president.

9. Many of their traditional crafts were _____ and useful both for lifestyle and health in general.

10. He believes his public repudiation of the conference decision will _____ his standing as a leader.

Banked cloze

Fill in the blanks by selecting suitable words from the word bank. You may not use any of the words more than once.

A. devote	F. texture	K. revolution
B. distinct	G. monetary	L. undertake
C. assess	H. tough	M. visualize
D. occasion	I. attribute	N. mutual
E. insecurity	J. pursue	O. enhance

Successful fashion designers have a wide array of skills, including drawing, an eye for color and 1. _____, an ability to 2. _____ concepts in three dimensions, and the mechanical skills involved in sewing and cutting all types of fabrics.

3. _____ your skills and personality honestly before 4. _____ a career in fashion design. You may love clothes but clothing is only part of the story when 5. _____ fashion design. You'll also need excellent communication skills, a willingness to work very hard (often 24/7), a 6. _____ hide when criticized, an ability to cope with stress, openness to having many different clients and/or bosses, an acceptance that there will be loneliness or isolation on 7. _____ (depending on how you set up your design business or career) and an ability to be a self-disciplined self-starter.

Being a fashion designer is probably for you if: You want to 8. _____ your life to this career (it's your "vocation"), you don't mind uncertainty or 9. _____; you are willing to stand up for what you believe in; you have 10. _____ ideas about what is important in fashion, and you listen to clients well; you know the fashion industry inside out and you live, eat and breathe fashion.

Expressions in use

Fill in the blanks with the expressions given below. Change the form when necessary. Each expression can be used only once.

be viewed as	indispensable for	contribute to	associate with
summarize into	dependent on	add to	vary with

1. It is said that medical negligence has _____ her death.
2. The menu in the restaurant _____ the season.
3. Low oestrogen levels _____, in general, _____ stress.
4. To sum it up, at least at this stage, it's unlikely that online learning will _____ the dominant form of education.
5. A good dictionary is _____ learning a foreign language.
6. We can _____ all the orientations of curriculum evaluation _____ three: objective orientation, process orientation, and subjectivity orientation.
7. The festival is heavily _____ sponsorship for its success.
8. This latest incident will _____ the pressure on the White House.

◼ Translation

I. Translate the following paragraph into Chinese.

An haute couture fashion designer is responsible for designing individualized, customized clothing for elite clientele. Clients are taken one at a time, and are given undivided attention. Designs are conceptualized and constructed according to a client's exact measurements, style, preferences, and personality. Each piece is made by hand from beginning to end, from expensive and high-quality fabric, and sewn with extreme attention to detail by the best seamstresses and embroiderers in the world. Considering the amount of time, money, and skill needed to complete each piece, haute couture garments typically have no price tag.

Although individualized, customized clothing for elite clientele is still going strong, today's haute couture fashion designs that are seen on runways are not particularly made to be sold or a main source of income. Rather, they are mostly for show and to further publicity, as well as perception and understanding of brand image. This brand image adds allure to a designer's prêt-à-porter (ready-to-wear) clothing line and to related high-end products such as shoes, purses, and perfumes.

II. Translate the following paragraph into English.

依据行业领域和责任的不同，纺织品设计师可以分为初级助理设计师和高级设计师。初级助理设计师的工作主要是辅助设计团队，高级设计师则是监管设计团队的工作。通常来说，纺织品设计师的工作是设计原样，并按照工厂设定的任务将它完善为成品。要想成为一名纺织品设计师，必须要具有强烈的时尚感，以及研究能力和设计技能。选修相关课程、拿到相关学位有助于学生实现自己的愿望。

Paragraph Writing

How to develop an example essay

In this unit, you will learn how to write an example essay. In an example essay, you support your point by illustrating it with examples. Vivid examples light up abstract ideas and make them clear, interesting, memorable, or convincing.

How many examples you use in an example essay depends on the topic. Some topics require numerous examples; others can be effectively developed with three or four extended examples. For instance, the thesis statement "San Francisco has some of the most unusual sights in California" does not require numerous examples, so three or four examples would be enough. Also, the examples you use to develop the thesis statement should be representative—fairly support your thesis.

A successful example essay depends on the following two guidelines:

· **A wise selection of sufficient examples which are specific and typical, interesting and relevant.** The examples may consist of either personal experiences or second-hand information from reliable sources.

· **An expert arrangement of these examples.** Similar or related examples should be grouped together and arranged according to the order of time, space, or importance.

To make examples specific and impressive, you need to use descriptive words and direct speeches. In addition, to achieve coherence in your example essay, you need to pay attention to the order of examples and proper use of transitions. The examples in an example essay can be organized according to time, familiarity and importance, just as the organization of the details in an expository essay. Meanwhile, body paragraphs in an example essay must be connected so that they can flow smoothly. In other words, the shift from one example to the next shouldn't be abrupt so that the reader understands clearly the progression of thought.

Transitions to introduce examples include:

One example of ... is ...

First, consider ...

To begin with, ...

Another example of ... is ...

An additional example is ...

Second / Next, consider ...

Still another example of ... is ...

Third / Finally, consider ...

The most significant example of ... is ...

Read the sample essay and see how the examples are developed and organized.

Topic

Gender barriers

Introduction

Thesis statement: Female managers today are still facing many challenges, especially gender barriers.

Body

Example 1: A double standard of the behavior of men and women

Example 2: A stereotyped "bimbo" title for women

Conclusion

Female managers can be made to feel more insecure at work because of gender barriers.

Sample essay

Throughout history, women have been hard workers who place high value on achievement. They often judge their personal worth based upon their accomplishments rather than how much money they make. At the management level, women are also considered better managers than men because of their unique working styles. However, female managers today are still facing many challenges, especially gender barriers.

To begin with, a double standard occurs when the behaviors of men and women are labeled. Like it or not, we still have some notions of what is an acceptable female behavior. Those women who work extra hard, make no allowances for failure, and don't believe in weaknesses are considered cold-hearted. When a woman shows strength of character in a meeting, for instance, she may be seen as "aggressive" whereas her male counterpart is seen as "assertive".

An additional example of gender barriers for women in the workplace is the "bimbo" title, which is used to describe a woman who got her position by playing a game of her female tricks. For this kind of women, they are pretty but not necessarily intelligent. So, the general concept is that women are incompetent and that they remain in the position only by means of their charm on the male bosses. But it could all be just a matter of our own perceptions of what's normal for a man and a woman.

To conclude, female managers can be made to feel more insecure at work because of gender barriers. They have to face a double standard of the behavior of men and women; they also worry about some negative stereotypes of women.

Write an essay of no less than 150 words on the topic "My troublesome apartment". You can follow the outline given below.

Topic

My troublesome apartment

Introduction

Thesis statement: My apartment has given me nothing but headaches.

Body

Example 1: My landlord has been uncooperative.

Example 2: I also have problems with the janitor (大楼管理员).

Example 3: The worst trouble is with the neighbors who live above.

Conclusion

My apartment is surrounded by all these troubles, which make me think about moving out.

How to Be a Fashion Designer

1   Before **addressing** how to become a fashion designer, it would be useful to **dispel** a few **myths** or **misconceptions**. Usually, people think that the profession of a fashion designer is based on creativity and imagination and anyone can become a fashion designer with these qualities. However, this is not true. No doubt that creativity and **innovative** ideas are the basic requirement for fashion designing, but one needs to **polish** these abilities to create masterpieces.

2   To become a fashion designer, the first step one must take is to join a fashion design school or college. Fashion design schools are the institutions providing **prospective** designers with the necessary training. The fashion design schools give the basic knowledge of design. While they are not required, many fashion designers will have an associate or **bachelor's** degree in fashion design or a related field. As a fashion design major, one will take classes in color, textiles, sewing and **tailoring**, pattern making, fashion history, and computer-aided design (CAD) and learn about different types of clothing such as menswear or **footwear**.

3   In the fashion design schools, the instructors teach the students to draw the designs, sew, and acquire **fundamentals** of colors. The training also enables one to become more confident about his skills and **portray** them with **dexterity** at workplace. Most of the time, people practice the art of sewing and designing at home. They do not attend any fashion design school or college and prefer to learn at home. **There is no point** denying the fact that practice makes a man perfect, but fashion design institutions help one shape the abilities. They give a direction and make one more focused. Another reason why joining a fashion design school or college is the key to a successful career in designing is that, these days, the

companies and other employers demand the professionals with degree or certificate that prove their **expertise** in the field of fashion design. The certificates prove that one has undertaken a formal training and **are apt for** the job. Therefore, attending a fashion design institute becomes the **prime** requirement to be a successful fashion designer.

4   Another effective way that shows how to become a fashion designer is to work as an **intern** or as an **apprentice** with a leading designer or a fashion house. **Internship** is a small period when one works as a trainee and observes the requirements of the business he is going to enter. During the education, if one has worked as a trainee, the employers prefer to appoint him as he **is acquainted with** the job type.

This job requires designers to be able to perform duties that include the following:

· Draw, design, develop new clothing and **accessory** samples, create and turn in **documentation** worksheets.

· Participate in meetings to discuss designs and line development, present and review line and concepts regularly.

· Identify new opportunities relevant to the firm's customer base through market research.

· Manage new design and fit styles while also maintaining **corporate** standards for **bulk** production.

· Cost out all of the components needed for a **garment's** production.

· Assist a design team with communicating to **vendors** on design, production, and other issues.

5   The design process from initial design concept to final production takes 18 to 24 months. The first step in creating a design is researching current fashion and making predictions of future trends. Some designers conduct their own research, while others rely on trend reports published by fashion industry trade groups. Trend reports indicate what styles, colors and fabrics will be popular for a particular season in the future. Textile manufacturers use these trend reports to begin designing fabrics and patterns while

fashion designers **sketch** preliminary design. Designers then visit manufacturers or trade shows to **procure** samples of fabrics and decide which fabrics to use.

6   Once designs and fabrics are chosen, a **prototype** of the article using cheaper materials is created and then tried on a model to see what adjustments need to be made. This also helps designers to narrow their choices of designs to offer for sale. After the final adjustments and selections have been made, samples of the article using the actual materials are sewn and then marketed to clothing retailers. Many designs are shown at fashion trade a few times a year. **Retailers** at the shows place orders for certain items, which are then manufactured and distributed to stores.

7   Computer-aided design is increasingly being used in the fashion design industry. Although most designers initially sketch designs by hand, a growing number also translate these hand sketches to the computer. CAD allows designers to view designs of clothing on virtual models and in various colors and shapes, thus saving time by requiring fewer adjustment prototypes and samples later.

8   Depending on the size of their design firm and their experience, fashion designers may have varying levels of involvement in different aspects of design and production. In large design firms, fashion designers are often the leading designers who are responsible for creating the designs, choosing the colors and fabrics, and **overseeing** technical designers who **turn** the designs **into** a final product. They **are responsible for** creating the prototypes and patterns and work with the manufacturers and suppliers during the production stages. Large design houses also employ their own patternmakers, tailors, and sewers who create the master patterns for the design and sew the prototypes and samples. Designers working in small firms, or those new to the job, usually perform most of the technical, patternmaking and sewing tasks, in addition to designing the clothing.

9   Fashion designers working for **apparel wholesalers** or manufacturers create designs for the mass market. These designs are manufactured in

various sizes and colors. A small number of high-fashion (haute couture) designers are self-employed and create customized designs for individual clients, usually at very high prices. Other high-fashion designers sell their designs in their own retail stores or **cater to specialty** stores or high-fashion department stores. These designers create a mixture of original garments and those that follow established fashion trends.

10   Some fashion designers **specialize in** costume design for performing arts, motion picture, and television productions. The work of costume designers is similar to that of other fashion designers. Costume designers, however, perform extensive research on the styles worn during the period in which the performances takes place, or they work with directors to select and create appropriate **attire**. They make sketches of designs, select fabric and other materials, and oversee the production of the costumes. They also must keep the spending within the costume budget for the particular production item.

11   If they are hoping to become the next top designer, the chances are **slim** due to the competitiveness of the industry. Although some designers become household names, most remain unknown to the general public and **anonymously** create the designs behind well-known brands and lesser-known labels.

New words and phrases

address /əˈdres/

n. ① details of where sb lives or works and where letters, etc. can be sent 住址；地址；通信处 ② [C] a formal speech delivered to an audience 演说；演讲，致辞

v. ① [usually passive] ~ **sth (to sb/sth)** to write on an envelope, etc. the name and address of the person, company, etc. that you are sending it to by mail 写（收信人）姓名地址；致函 ② to make a formal speech to a group of people 演说；演讲 ③ ~ **sb**|~ **sth to sb** (*formal*) to say sth directly to sb 向……说话

dispel /dɪˈspel/   *v.* to make sth, especially a feeling or belief, go away or disappear 驱散；消除（尤指感觉或信仰）

myth /mɪθ/   [C, U] ① a story from ancient times, especially one that was told to explain natural events or to describe the early history of a people; this type of story 神话；神话故事 ② sth that many people believe but that does not exist or is false 虚构的东西；荒诞的说法；不真实的事

misconception /ˌmɪskənˈsepʃn/   *n.* [C, U] ~ (about sth) a belief or an idea that is not based on correct information, or that is not understood by people 错误认识；误解

innovative /ˈɪnəveɪtɪv/   *a.* introducing or using new ideas, ways of doing sth, etc. 引进新思想的；采用新方法的；革新的；创新的

polish /ˈpɒlɪʃ/   *n.* ① [U, C] a substance used when rubbing a surface to make it smooth and shiny 擦光剂；上光剂；亮光剂 ② [U] a high quality of performance achieved with great skill（表演的）完美；娴熟；精湛 ③ [U] high standards of behavior; being polite 文雅；优雅；品味；礼貌

v. ① ~ sth (up) (with sth) to make sth smooth and shiny by rubbing it with a cloth, often with polish on it 擦光；磨光 ② ~ sth (up) to make changes to sth in order to improve it 修改；润饰；润色

prospective /prəˈspektɪv/   *n.* [usually before noun] ① expected to do sth or to become sth 有望的；可能的；预期的；潜在的 ② expected to happen soon 即将发生的；即将来临的

bachelor /ˈbætʃələ(r)/   *n.* ① a man who has never been married 未婚男子；单身汉 ② (usually Bachelor) a person who has a Bachelor's degree (= a first university degree) 学士 *a Bachelor of Arts/Engineering/Science* 文学士；工程学士；理学士

tailoring /ˈteɪlərɪŋ/   *n.* [U] ① the style or the way in which a suit, jacket, etc. is made 裁剪式样；裁缝手艺 ② the job of making men's clothes（男装）裁缝业；成衣活

footwear /ˈfʊtweə(r)/   *n.* [U] things that people wear on their feet, for example shoes and boots 鞋类（如鞋和靴）

fundamental /ˌfʌndəˈmentl/   *a.* ① serious and very important; affecting the most central and important parts of sth 十分重大的；根本的 ② ~ **(to sth)** central; forming the necessary basis of sth 基础的；基本的 ③ [only before noun] forming the source or base from which everything else is made; not able to be divided any further 基本的；不能再分的

n. [usually pl.] a basic rule or principle; an essential part 基本规律；根本法则；基本原理；基础

portray /pɔːˈtreɪ/   *v.* ① to show sb/sth in a picture; to describe sb/sth in a piece of writing 描绘；描画；描写 ② ~ **sb/ sth (as sb/sth)** to describe or show sb/sth in a particular way, especially when this does not give a complete or accurate impression of what they are like 将……描写成；给人以某种印象；表现 ③ to act a particular role in a film/movie or play 扮演（某角色）

dexterity /dekˈsterəti/   *n.* [U] skill in using your hands or your mind（手）灵巧；熟练；（思维）敏捷；灵活

expertise /ˌekspɜːˈtiːz/   *n.* [U] ~ **(in sth/in doing sth)** expert knowledge or skill in a particular subject, activity or job 专门知识；专门技能；专长

prime /praɪm/   *a.* ① [only before noun] main; most important; basic 主要的；首要的；基本的 ② of the best quality; excellent 优质的；上乘的；优异的

n. [sing.] the time in your life when you are strongest or most successful 盛年；年富力强的时期；鼎盛时期

intern /'ɪntɜːn/

v. [often passive] ~ **sb (in sth)** to put sb in prison during a war or for political reasons, although they have not been charged with a crime （战争期间或由于政治原因未经审讯）拘留；禁闭；关押

n. ① an advanced student of medicine, whose training is nearly finished and who is working in a hospital to get further practical experience 实习医生 ② a student or new graduate who is getting practical experience in a job, for example during the summer holiday/vacation 实习学生；毕业实习生

apprentice /ə'prentɪs/

n. a young person who works for an employer for a fixed period of time in order to learn the particular skills needed in their job 学徒；徒弟

v. [usually passive] ~ **sb (to sb) (as sth)** (*old fashioned*) to make sb an apprentice 使某人当（某人的）学徒

internship /'ɪntɜːnʃɪp/

n. ① a period of time during which a student or new graduate gets practical experience in a job, for example during the summer holiday/vacation （学生或毕业生的）实习期 ② a job that an advanced student of medicine, whose training is nearly finished, does in a hospital to get further practical experience 医科学生的实习工

accessory /ək'sesəri/

n. ① [usually pl.] an extra piece of equipment that is useful but not essential or that can be added to sth else as a decoration 附件；配件；附属物 ② [usually pl.] a thing that you can wear or carry that matches your clothes, for example a belt or a bag （衣服的）配饰

a. (*technica* 术语) not the most important when compared to others 辅助的；副的

documentation
/ˌdɒkjumenˈteɪʃn/
n. [U] ① the documents that are required for sth, or that give evidence or proof of sth 必备资料；证明文件 ② the act of recording sth in a document; the state of being recorded in a document 文件记载；文献记录；归档

corporate /ˈkɔ:pərət/
a. [only before noun] ① connected with a corporation 公司的 ② (*technical* 术语) forming a corporation 组成公司（或团体）的；法人的

bulk /bʌlk/
n. ① [sing.] **the ~ (of sth)** the main part of sth; most of sth 主体；大部分 ② [U] the (large) size or quantity of sth （大）体积；大（量）

v. to be the most important part of sth 是……的最重要部分

garment /ˈgɑ:mənt/
n. a piece of clothing 衣服

vendor /ˈvendə(r)/
n. ① someone who sells things such as newspapers, cigarettes, or food from a small stall or cart 小贩 ② a company or person that sells a product or service, especially one who sells to other companies that sell to the public 卖主

sketch /sketʃ/
n. ① a simple picture that is drawn quickly and does not have many details 素描；速写；草图 ② a short report or story that gives only basic details about sth 简报；速写；概述

v. ① to make a quick drawing of sb/sth 画素描；画速写 ② **~ sth (out)** to give a general description of sth, giving only the basic facts, outline 概述；简述

procure /prəˈkjʊə(r)/
v. **~ sth (for sb/sth)** (*formal*) to obtain sth, especially with difficulty （设法）获得；取得；得到

prototype /ˈprəʊtətaɪp/   *n.* ~ **(for/of sth)** the first design of sth from which other forms are copied or developed 原型；雏形；最初形态

retailer /ˈriːteɪlə(r)/   *n.* a person or business that sells goods to the public 零售商；零售店

oversee /ˌəʊvəˈsiː/   *v.* supervise; to watch sb/sth and make sure that a job or an activity is done correctly 监督；监视

apparel /əˈpærəl/   *n.* [U] ① (especially *NAmE*) clothing, when it is being sold in shops/stores（商店出售的）衣服；服装 ② (*old-fashioned*) (*formal*) clothes, particularly those worn on a formal occasion（尤指正式场合穿的）衣服；服装

wholesaler /ˈhəʊlseɪlə(r)/   *n.* a person whose business is buying large quantities of goods and selling them in smaller amounts, for example, to stores 批发商

cater /ˈkeɪtə(r)/   *v.* ~ **(for sb/sth)** to provide food and drinks for a social event（为社交活动）提供饮食；承办酒席

speciality /ˌspeʃiˈæləti/   *n.* (*pl.* **-ies**) ① a type of food or product that a restaurant or place is famous for because it is so good 特产；特色菜 ② an area of work or study that sb gives most of their attention to and knows a lot about; sth that sb is good at 专业；专长

specialize /ˈspeʃəlaɪz/   *v.* ~ **(in sth)** to become an expert in a particular area of work, study or business; to spend more time on one area of work, etc. than on others 专门研究（或从事）；专攻

attire /əˈtaɪə(r)/   *n.* [U] (*formal*) clothes 服装；衣服

slim /slɪm/   *a.* ① (*approving*) (of a person 人) thin, in a way that is attractive 苗条的；纤细的 ② not as big as you would like or expect 微薄的；不足的；少的；小的

v. ① (usually used in the progressive tenses 通常用于进行时) to try to become thinner, for example by eating less （靠节食等）变苗条；减肥 ◇ **slim down** to make a company or an organization smaller, by reducing the number of jobs in it; to be made smaller in this way 精简（机构）；裁减（人员）；减少（岗位）

anonymously /əˈnɒnɪməsli/	*ad.* without giving a name 匿名地；不署名地
there is no point doing …	there is no need to do sth 做某事没有意义或作用
be apt for	are suitable or appropriate for… 适合
be acquainted with	be familiar with sth, having read, seen or experienced it 熟悉；了解
turn … into …	① to change and become sb or sth different 把……变成…… ② to make sth change the direction it is moving in (使) 改变方向
be responsible for	① being the cause of sth 作为原因；成为起因 ② having the job or duty of doing sth 有责任；负责
cater to	to provide the things that a particular type or person wants, especially things that you do not approve of 满足需要；迎合
specialize in	to become an expert in a particular area of work, study or business 专门研究；专攻

◼ Reading Comprehension

Understanding the text

Answer the following questions.

1. What is true when it comes to the profession of a fashion designer?

  A. The profession of a fashion designer is based on creativity and imagination.

B. Anyone can be a fashion designer.

C. Creativity and innovative ideas are the basic requirements for fashion design.

D. One needs to polish qualities like imagination and innovation to create masterpieces.

2. Why is it the first step to join a fashion design school or college for those who want to be successful?

A. Because it is required by the profession.

B. Because they prefer to learn the necessary training in the college rather than at home.

C. Because universities provide them with the basic knowledge of design and university degrees prove their expertise in the fashion field.

D. Because joining the university is the key to a successful career.

3. How do people become fashion designers?

A. To join a fashion design school or college.

B. To acquire such basic requirements as creativity and innovation.

C. To work as an intern or an apprentice.

D. Both A and C.

4. The fashion designers performed many duties expert for _____.

A. drawing, designing, and developing new clothing

B. discussing designs and line development and presenting them regularly

C. practicing sewing and designing at home

D. communicating with vendors on design, production and other issues

5. When it comes to creating a design, what should be done at first?

A. To research current fashion and make predictions about future trends.

B. To create custom designs for individual clients.

C. To follow established fashion trends.

D. To choose the colors and fabrics to cater to different clients.

6. How long would the design process take from initial design concept to final production?

A. Less than a week.

B. More than two days.

C. Between 18 and 24 hours.

D. No less than one day.

7. Why is computer-aided design increasingly being applied to the fashion design industry?

    A. Because it is a new trend.

    B. Because it saves time.

    C. Because it relieves the pressure of fashion designers.

    D. Because it assists the work of fashion designers.

8. What is the purpose of writing this passage?

    A. To entertain readers.

    B. To inform readers of the duties of fashion designers.

    C. To argue the competitiveness of the fashion industry.

    D. To advise readers the successful way to be a fashion designer.

Critical thinking

Work in pairs and discuss the following questions.

1. What benefits can a person get from working as an intern or as an apprentice with a leading designer?

2. What qualifications are required for a fashion designer?

3. Why do so many people want to become fashion designers even if the chances are slim?

■ Language Enhancement

Words in use

Fill in the blanks with the words given below. Change the form when necessary. Each word can be used only once.

dispel	polish	oversee	procure	expertise
anonymous	specialize	prime	dexterity	sketch

1. The location of these beaches makes them _____ sites for development.

2. Nancy was not an accountant and didn't have the _____ to verify all of the

financial details.

3. The shop _____ in hand-made chocolates and delicious cookies.

4. Queen's speech on New Year's eve _____ any fears about her health.

5. To secure the quality of the project, we'd better use a surveyor or architect to _____ and inspect the different stages of the work.

6. The money was donated by a local businessman who wishes to be _____.

7. It remained very difficult to _____ food, fuel, and other daily necessities in poverty areas in this country in 1980s.

8. Caption _____ the story briefly, telling the facts just as they had happened.

9. You need manual _____ to be good at playing guitar.

10. The statement had been carefully _____ and checked before it was released.

Expressions in use

Fill in the blanks with the expressions given below. Change the form when necessary. Each expression can be used only once.

be apt for/to	be acquainted with	specialize in
be responsible for	in addition to	cater to
turn … into…	there is no denying …	

1. Many individuals _____ more _____ seek informal support from family and friends, though they may not be sufficient to prevent their long-term distress.

2. _____ that black tea is not enjoyed for everyone.

3. The party is betting that the presidential race will _____ a battle for younger voters.

4. In the contemporary western world, some rapidly changing styles _____ a desire for novelty and individualism.

5. The best leaders have always understood that they have to _____ all elements of their organizations.

6. _____ earthquakes and tsunamis, global warming is a major concern for the entire planet nowadays.

7. The storm is thought to _____ as many as 40 deaths.

8. Navy Seals _____ hostage rescue, counter-terrorism, unconventional warfare—activities that sound like a great video game.

Sentence structure

I. Complete the following sentences by translating the Chinese into English, using "there is no/little doubt +that ..." structure.

Model: _____ (毫无疑问, 创意和创新思想是时装设计的基本要求), but one needs to polish these abilities to create masterpieces.

→There is no doubt that creativity and innovative ideas are the basic requirement for fashion designing, but one needs to polish these abilities to create masterpieces.

1. _____ (毫无疑问, 上大学能让年轻人接触到新思想) and helps promote their critical thinking.

2. _____ (毫无疑问, 计算机犯罪是一个很严重的问题), so people think that all hackers need to be punished for their actions.

3. Though the boy worked very hard and finally successfully passed the exam, _____ (毫无疑问, 过度劳累导致了他的病情).

II. Rewrite the following sentences by using "there is no point doing that ..."

Model: It is true that practice makes a man perfect, but fashion design institutions help one shape the abilities.

→There is no point denying the fact that practice makes a man perfect, but fashion design institutions help one shape the abilities.

1. Even though it is important for the students to have a deep understanding of the texts, it is of little meaning to read texts word for word from the beginning to the end.

_____

_____

_____.

2. As it is a matter of little importance to us, it doesn't make much sense to argue who is the winner of the game.

_____

_____

_____.

3. It is unreasonable to compel children to obey their parents; instead, we should respect their own decisions and individualism.

_____

_____

_____.

Extensive Reading

Coco Chanel—The Most Beautiful and Successful Woman in the Fashion World

1   Named as one of the 100 most influential people of the 20th century, legendary designer Coco Chanel was a pioneer in the world of fashion. Her kind of fashion was inspired by menswear and she believed in simple yet highly expensive designs. Not only did she make "black" the black it is today, but she also created the wardrobe staple that none of us can live without the little black dress!

2   Coco Chanel held sway over haute couture for six decades. Chanel's collections were casual clothes that were elegantly styled. In 1954 she

introduced the collarless cardigan jacket and bell-bottom pants. Chanel was famous for her quilted purses, the "little black dress," and her perfume—Chanel No.5—possibly the most famous fragrance of all time.

3 Gabrielle Coco Chanel was born on August 19, 1883, in the hospital of the French village of Saumur. She was only twelve when her mother died; her father, a travelling wine peddler, then took her and her sister to convent, consigning them into the care of the nuns and disappearing to America forever.

4 She discovered the elements of what was to make her a great designer, that mix of darkness and light, white and black. Pearls, which are so much a part of Chanel's style, seem very much reminiscent of the rosary beads the nun wore, the chains around the nun's waist. There are still tiny chains sewn in the bottom of every Chanel jacket.

5 At eighteen, Chanel left the convent to strike out on her own. She worked as an assistant at a tailoring shop but expected to be a music hall performer. Later, she opened a small shop to sell fashionable hats to ladies. As her business grew, she became a dressmaker and created her own style in clothing. The Chanel style has everything to do with elegance, comfort, ease and practicality. Among her design innovations, all trademarks were the use of jersey and fake jewelry, consisting of pearl ropes and colored crystal hanging around her neck. In 1926, she introduced the little black dress and for the first time this color had been in favor (except for mourning) since Spain of the 15th century. *Vogue* predicted it would not only be the dress of the decade, but for the first time, used the word "ford" in describing it. In the fashion industry, the word "ford" means that the style will be worn by millions, i.e., everybody's dress.

6 Chanel bobbed her hair, and so did most of the western world; she dared to wear pants—copied somewhat after a sailors; she tanned her skin in the Mediterranean sun. But perhaps her most enduring legacy is the Chanel suit—a simple, cropped jacket usually bound with brad, soutache, or

ribbons, open and unbuttoned, with a soft, straight, or pleated skirt. This style was as fresh and new in 1925 as in the late 1990s, copied by almost every designer in the world at one time or another.

7   In the 1950s and 1960s, Chanel was at the top of her fame. She started dressing everybody from Brigitte Bardot, Jane Fonda, Elizabeth Taylor and Marilyn Monroe. Chanel also dressed Jackie Kennedy.

8   At the beginning of World War II, she closed her house, but even by that time, she was an extremely rich woman. She had no plans to reopen after the war, but it is said she was so competitive that the success of Dior inspired her to go back into business. She reopened in 1954, within a few months, once again, selling like hot cakes. It is hard to believe that a 71-year-old woman could reconquer the tough fashion world, but such was the case. Until her death in 1971, at the age of 87, she kept at her drawing board, turning out one collection after another. Her last collection was as welcomed as her first. She captured the spirit of a century. Perhaps her own words sum her life up best: speaking in the third person, she said "Chanel, above all else, is a style. Fashion, you see, goes out of fashion. Style, never."

9   Coco Chanel died on January 10, 1971. Hundreds crowded together at the Church of the Madeleine to bid farewell to the fashion icon. In tribute, many of the mourners wore Chanel suits.

10   Menkes, fashion editor of *The International Herald Tribune*, adds, "Chanel's secret is that clarity, that modernity, it's just the signage of the word "C", and somehow to me it looks very forward thinking, very modern, very dynamic. Certainly the brand was very well managed since she herself passed away. Karl Lagerfeld has certainly rejuvenated Chanel and continues to do so." Menkes says that's why Chanel is still an iconic brand. The designer lives on as inspiration itself.

11   Coco Chanel believed that if function came first, beauty would follow, and then fashion. This woman with a quick mind brought together looking

smart and being smart. George Bernard Shaw referred to Chanel as one of the two most important women living in the world (the other being Marie Curie).

Unit 4

Asian Traditional Costumes and Culture

Although culture is not as extensive as civilization, it has a more profound relationship with the happiness, anger, sorrow and joy of everyone in each country.

—*Moriya Masanori*

Culture without force will become an extinct culture tomorrow.

—*Winston Churchill*

Pre-Reading Activities

1. Listen to the recording and answer the following questions.

(1) Name the Four Ancient Civilizations respectively.

(2) Cite at least 3 descriptive words to depict China according to the recording.

2. Listen to the recording again and fill in the blanks based on what you hear.

China is one of the Four Ancient (1) _____ (alongside Babylon, India and Egypt). It (2) _____ a vast and varied geographic expanse, 3,600 years of written history, as well as a rich and profound culture. Chinese culture is (3) _____ and unique, yet harmoniously (4) _____ an invaluable asset to the world.

Clothing in China has always varied by region, and over the years it has also changed rather (5) _____. During the imperial era, fashion was highly distinguished based on the social status of an individual. Silk was a (6) _____ left for the extremely wealthy of the society. Silk was not only a primary material in the design of outfits but was also a (7) _____ trade commodity for the country. The Silk Road was a vital network of trading routes that date back as early as the Han dynasty. These trade routes connected Asia with the Middle East, East Africa, and southern Europe. China has remained one of the biggest (8) _____ of silk to date.

3. Discuss the following questions with your partner.

(1) The Silk Road was a vital network of trading routes that date back as early as 207 BC, what do you know about the history of the Silk Road ?

(2) Chinese culture is distinctive and predominant. Can you give a brief introduction to some well-known traditional Chinese clothes types, namely, Hanfu, Tang suit, and qipao, etc.

China's Ancient Costumes and Culture

1   China **boasts** a long history of costumes **dating back to** the earliest primitive society. With the advancement of China's social culture and civilization, people demanded clothes to be more than a covering of the body to keep warm. People increasingly focused on the **pursuit** of beauty, which contributed to the development of Chinese costumes. People **integrated** customs, beauty, color patterns, culture and religion into the costumes, which led to formation of the **distinctive** and profound Chinese clothing culture.

2   From time **immemorial**, human beings used animal hides, leaves, grass and feathers to keep warm. These were the prototypes for the original clothes. Dating back to 10000 BCE, people entered into the Neolithic Age. Chinese ancestors gradually acquired textile technology **ushering** in a civilized society where people had real sense of "clothes ".

3   With time passing by, the ancient people discovered hemp. Using this material, they wore dress. Ever since then the primitive textile industry rose. The ancient people started to raise silkworm to **spin** and make silk in the late primitive society. As recorded in "Xici Xia" of *Yijing*, during the **reign** of Yellow Emperor , Yao and Shun , the system of China's costumes and accessories had been initially established. It had become a vehicle of social institution and continued to use in following several thousands years of Chinese history.

4   In Xia and Shang dynasties, a **rigid hierarchy** of China's costumes became more distinct. The hierarchical system of "**six-crown and six-gown**" was set up. It demonstrated that costume had been an **attribute** of power and hierarchy.

5   During the Warring States Period, King Wuling of Zhao State performed a profound reform "wearing Hu dress and shooting on horseback", that is, learning to ride and **archery** from the western ethnic minorities and starting to wear their costumes. Following this trend, people started to wear **formfitting** and light clothes.

6   The idea of fashion reached a new height during the Spring and Autumn and the Warring States periods, when wars broke out frequently and the various states **spared no effort** to enhance their strength. The different styles of clothes showed people's positions and the states they came from.

7   The Qin and Han dynasties **witnessed** the **unification** of territory as well as written language. Qin Shihuang, the First Emperor of the Qin dynasty, established many social systems, including one for **uniforms** to **distinguish** people's ranks and social positions. China's complete **code** of costume and **trappings** was established in the Han dynasty. The yarn-dyeing, embroidering and metal-processing technologies developed rapidly in the period, **spurring** changes in costume and **adornments**. By the Han dynasty, men wore gown-style dress which was taken as formal and noble attire, while women wore *Shenyi*, which **was composed of ramie**, or linen, fabric that needed to be **bleached**.

8   Loose style costume **prevailed** in Wei and Jin dynasties. Men wore loose clothes with large sleeves. Women's typical costume was *Za Ju Chui Shao* dress, which featured wide top and **tapering** bottom.

9   China underwent economic prosperity and political **stability** during Sui and Tang dynasties. Costumes at that period of time were **dignified** and graceful, especially for women's costumes, and they were rich in color and style. They **tended to** be open and romantic. Women in the Tang dynasty usually wore short coat and long skirt with high-tied waist. Costumes of the Song dynasty were **shifted to** be simple and natural. Men mostly wore clothes with loose sleeves and round collar. Women's clothes for upper part of the body included short coat, *Ao,* **underlinen,** *Beizi* and **half-sleeve**

garments. Clothes for lower part of the body were skirt.

10   Costumes of Han were resumed in the Ming dynasty. Strict rules for costumes were set up. Social institutions were stressed. Ranks for emperor's and officials' costumes were **formulated**. Formal attire and casual wear were **differentiated**.

11   Casual wear appeared during the Song dynasty, and clothes were simple and elegant. Clothes were luxurious for upper class yet simple and **unadorned** in design. **Dramatic** changes took place during the Ming dynasty. A new conception **came into being** in clothing design, with no limitation to one style and advocating natural beauty, thus bringing **vigor** and **vitality** to the clothing culture.

12   In the Qing dynasty, distinctive clothing feature and heavy-coated style were formed. During the Qing dynasty, clothes became elegant, **poised** and glorious. The entire world witnessed dramatic changes such as **the Renaissance** in Italy and **Columbus's** discovery of the American Continent, but the changes did not affect traditional Chinese clothing.

13   The Chinese style of clothing **emanated** certain messages in society in the ancient feudal period. Garments were considered as more than just fabrics to cover one's body. Everyone could easily figure people's social status and rank out from what they wore **on a daily basis**. This **applied for** the upper-class. For instance, the color yellow was strictly assigned to the Emperor. The dragon **emblem** on traditional Chinese **imperial** dress served as an exclusive affirmation of their power.

14   Traditional clothing played a vital part of Chinese civilization. It has a significant role in the country's culture and history. Today, the culture of wearing traditional attires is still very much alive. People wear them during ceremonies, festivals or religious occasions.

Notes

the Renaissance   the period in Europe during the 14th, 15th and 16th centuries when people became interested in the ideas and culture of ancient Greece and Rome and used these influences in their own art, literature, etc. 文艺复兴

Columbus 哥伦布（意大利航海家）

XiCi Xia of *Yijng* 《易经·系辞下》

six-crown and six-gown 六服（古代天子及诸侯、卿大夫的六种服色）

Shenyi 深衣（古代礼服）

Za Ju Chui Shao 魏晋杂裾垂髾服

Ao 袄

Beizi 褙子

New words and phrases

boast /bəʊst/	*v.* ① (not used in the progressive tenses 不用于进行时) to have sth that is impressive and that you can be proud of 有（值得自豪的东西） ② **~ (about/of sth)** to talk with too much pride about sth that you have or can do 自夸；自吹自擂
pursuit /pəˈsjuːt/	*n.* ① [U] the act of looking for or trying to find sth 追求；寻找 ② [C, usually pl.] sth that you give your time and energy to, that you do as a hobby 事业；消遣；爱好
integrate /ˈɪntɪɡreɪt/	*v.* ① **~ (A) (into/with B)** | **~ A and B** to combine two or more things so that they work together; to combine with sth else in this way （使）合并；成为一体 ② **~ (sb) (into/with sth)** to become or make sb become accepted as a member of a social group, especially when they come from a different culture （使）加入；融入群体
distinctive /dɪˈstɪŋktɪv/	*a.* having a characteristic that makes sth different and easily noticed 独特的；特别的；有特色的

immemorial /ˌɪməˈmɔːriəl/   *a.* beyond the limits of memory or tradition or recorded history 古老的；远古的；无法追忆的

usher /ˈʌʃə/   *v.* to take or show sb where they should go 把……引往；引导；引领

spin /spɪn/   *v.* ① ~ **(A into B)** | ~ **(B from A)** to make thread from wool, cotton, silk, etc. by twisting it 纺（线）;纺（纱） ② ~ **(sth) (round/around)** to turn round and round quickly; to make sth do this （使）快速旋转

reign /reɪn/   *n.* the period during which a king, queen, emperor, etc. rules 君主统治时期

dignified /ˈdɪɡnɪfaɪd/   *a.* calm and serious and deserving respect 庄重的；庄严的；有尊严的

rigid /ˈrɪdʒɪd/   *a.* ① (of rules, methods, etc.) very strict and difficult to be changed or varied （规则、规章或制度、法律）死板的；僵硬的；过于严格的 ② (of a person 人) not willing to change their ideas or behavior 固执的；僵化的；一成不变的

hierarchy /ˈhaɪərɑːki/   *n.* ① [C, U] a system, especially in a society or an organization, in which people are organized into different levels of importance from highest to lowest 等级制度（尤指社会或组织） ② [C] (*formal*) a system that ideas or beliefs can be arranged into 层次体系

attribute /əˈtrɪbjuːt/   *n.* a quality or feature of sb/sth 属性；性质；特征

v. ① ~ **sth to sth** to say or believe that sth is the result of a particular thing 把……归因于；认为……是由于 ② ~ **sth (to sb)** to say or believe that sb is responsible for doing sth, especially for saying, writing or painting sth 认为是……所为（或说、写、作）

archery /ˈɑːtʃəri/   *n.* [U] the art or sport of shooting arrows with a bow 射箭术；射箭运动

formfitting /'fɔ:mˌfɪtɪŋ/	*a.* fitting someone's body closely 合身的
witness /'wɪtnəs/	*v.* to be the place, period, organization, etc. in which particular events take place 是发生……的地点（或时间、组织等）；见证
unification /ˌjuːnɪfɪ'keɪʃn/	*n.* the process by which two or more countries join together and become one country 统一
uniform /'juːnɪfɔːm/	*n.* [C, U] the special set of clothes worn by all members of an organization or a group at work, or by children at school 制服；校服
	a. the same in all parts and at all times 一致的；统一的；一律的
distinguish /dɪ'stɪŋgwɪʃ/	*v.* ① ~ **(between) A and B | ~ A from B** to recognize the difference between two people or things 区分；辨别；分清 ② (not used in the progressive tenses 不用于进行时) to be able to see or hear sth 看清；认出；听出
code /kəʊd/	*n.* ① [C] a system of laws or written rules that state how people in an institution or a country should behave 法典；法规 ② [C, U] (often in compounds 常构成复合词) a system of words, letters, numbers or symbols that represent a message or record information secretly or in a shorter form 密码；暗码；电码；代码
trappings /'træpɪŋz/	*n.* [pl.] ~ (of sth) (*formal, especially disapproving*) the possessions, clothes, etc. that are connected with a particular situation, job or social position （与某一处境、职业或社会地位有关的）身外之物；标志；服装
spur /spɜː(r)/	① to make sth happen faster or sooner 促进；加速；刺激（某事发生）② ~ **sb/sth (on) (to sth/to do sth)** to encourage sb to do sth or to encourage them to try harder to achieve sth 鞭策；激励；刺激；鼓舞

adornment /əˈdɔ:nmənt/   *n.* sth that is used to make a person or thing more beautiful 装饰品

ramie /ˈræmɪ/   *n.* a woody urticaceous shrub of Asia, Boehmeria nivea, having broad leaves and a stem that yields a flaxlike fibre 苎麻纤维

bleach /bli:tʃ/   *v.* to make sth white or pale by a chemical process or by the effect of light from the sun; to become white or pale in this way （使）变白；漂白；晒白；褪色

prevail /prɪˈveɪl/   ① ~ **(in/among sth)** to exist or be very common at a particular time or in a particular place 普遍存在；盛行；流行 ② ~ **(against/over sth)** (*formal*) (of ideas, opinions, etc. 思想、观点等) to be accepted, especially after a struggle or an argument 被接受；战胜；压倒

taper /ˈteɪpə(r)/   *v.* to become gradually narrower; to make sth become gradually narrower （使）逐渐变窄

stability /stəˈbɪləti/   *n.* [U] the quality or state of being steady and not changing or being disturbed in any way (= the quality of being stable) 稳定（性）；稳固（性）

underlinen /ˈʌndəˌlɪnən/   *n.* underclothes, esp. when made of linen (亚麻) 内衣

regime /reɪˈʒi:m/   *n.* a method or system of government, especially one that has not been elected in a fair way （尤指未通过公正选举的）统治方式；统治制度；政权；政体

formulate /ˈfɔ:mjuleɪt/   *v.* to create or prepare sth carefully, giving particular attention to the details 制订；规划；构想；准备

differentiate /ˌdɪfəˈrenʃieɪt/   *v.* ~ **(between) A and B** | ~ **A (from B)** to recognize or show that two things are not the same 区分；区别；辨别

unadorned /ˌʌnəˈdɔ:nd/   *a.* (*formal*) without any decoration 不加装饰的；简朴的

dramatic /drəˈmætɪk/	*a.* (of a change, an event, etc.) sudden, very great and often surprising（变化、事情等）突然的；巨大的；令人吃惊的
vigor /ˈvɪɡə(r)/	*n.* energy, force or enthusiasm 精力；活力
vitality /vaɪˈtæləti/	*n.* having great energy and liveliness. 活力；生命力；热情
poised /pɔɪzd/	*a.* ① [not before noun] marked by balance or equilibrium and readiness for action 平衡；平稳 ② calmly, full of poise 泰然自若的；沉着的
emanate /ˈeməneɪt/	*v.* (*formal*) to produce or show sth 产生；表现；显示
emblem /ˈembləm/	*n.* a design or picture that represents a country or an organization（代表国家或组织的）徽章；标记；图案
imperial /ɪmˈpɪəriəl/	*a.* connected with an empire 帝国的；皇帝的
date back to	belong to an earlier time 追溯到
tend to	be likely to do sth 通常会……
shift to	change to another 变换，变动
come into being	come into existence 形成
on a daily basis	based on everyday activities 每天
apply for	make a formal request for sth 申请；请求
be composed of	be made from several parts 由……组成的

▪ Reading Comprehension

Understanding the text

Answer the following questions.

1. According to the author, what led to the development of Chinese costumes?

2. What kind of materials did the ancient humans use to keep warm?

3. When was the system of China's costumes and accessories been initially set up?

4. Why did fashion reach a new height during the Spring and Autumn and the Warring States periods?

5. What are the features of *Za Ju Chui Shao*?

6. What are the features of women's costumes during the Sui and Tang dynasties?

7. Illustrate the sentence "The Chinese style of clothing can emanate certain messages in the ancient China."

8. How do you understand the first sentence in the paragraph 14?

Critical thinking

The following chart will present you an overview of the features of China's ancient costumes evolution. Find out the key information and then complete it.

Ages	Features/materials of the costumes	illustration/sketch out
Primitive society	leaves, grass, feather and animal hide were used to cover the body	
Qin and Han dynasties	*Shenyi*:	
Wei and Jin dynasties		
Sui and Tang dynasties		
Song dynasty		
Yuan dynasty		
Qing dynasty		

Language Enhancement

Words in use

Fill in the blanks with the words given below. Change the form when necessary. Each word can be used only once.

boast	pursuit	integrate	rigid	attribute
prevail	unification	spur	emanate	differentiate

1. Several colleges in the study have _____ rules about student conduct.

2. It is hard for logic to _____ over emotion.

3. The former President of South Africa, Nelson Mandela pledged to _____ the whole country in a peaceful way.

4. Her difficult childhood _____ her on to succeed.

5. A child may not _____ between his imagination and the real world.

6. These new immigrants have made great efforts to _____ with the local community.

7. He _____ power and confidence especially in face of challenges.

8. The newly-built library will _____ the latest energy-saving technology.

9. A good teacher should possess patience, one of the most important _____ in teaching.

10. Children are encouraged to travel the world in order to _____ their dreams.

Banked cloze

Fill in the blanks by selecting suitable words from the word bank. You may not use any of the words more than once.

A. prestige	F. loose fitting	K. badge
B. formulate	G. preferred	L. adornment
C. prevailed	H. garments	M. cultivation
D. ethnic	I. immemorial	N. uniforms
E. identify	J. dyed	O. evolved

Ancient Chinese clothes were more than just 1. _____ to cover the body. They were the symbols of 2. _____ and the embodiment of culture, and essential to show a person's social status. The rich and poor dressed very differently in ancient China. Poor people wore hemp backpacks which were durable, 3. _____, and comfortable for working in the fields. On the other hand, rich people's clothes were made from silk, 4. _____ with specific colors and could be turned into fancy designs.

Lower-class people could be punished for wearing silk clothing. The color of a person's clothing was also a 5. _____ of identity in ancient China. In the Yuan dynasty, the use of cotton for clothing began to 6. _____.

Chinese fashion 7. _____ with time, but ancient Chinese clothing was very limited. 8. _____ and jewelry were not only part of fashion, but they were also symbols of social status. There were many rules about the wearing of jewelry. An individual could easily 9. _____ a person's social status by looking at their jewelry. Men used belt hooks or buckles, and women wore combs and hairpins. The ancient Chinese wore more silver than gold. They also used other materials like blue kingfisher feathers, blue gems, and glass.

The ancient Chinese 10. _____ jade over any other stone. They believed that jade had the human-like qualities of hardness, durability, and beauty. The early jade designs were simple, but they evolved over time. Amulets were commonly used as jewelry with the symbol of the dragon on them.

Expressions in use

Fill in the blanks with the expressions given below. Change the form when necessary. Each expression can be used only once.

| date back to | spare no effort | come into being | on a daily basis |
| apply for | tend to be | shift to | be composed of |

1. We should _____ to beautify and protect our environment.
2. Although a team may _____ knowledgeable people, they must learn new

ways of solving cross-functional problems.

3. A large-scale _____ online shopping has had a big impact on traditional shopping malls.

4. People _____ more aggressive when they're young and more conservative as they get older.

5. The tradition could _____ 300 BCE.

6. We can't afford too many organic products _____ since we struggle to make ends meet.

7. People can _____ citizenship after five years' residency in some countries.

8. Scientific researchers may be curious how the environment pollution _____.

▪ Translation

I. Translate the following paragraph into Chinese.

Perfume sachet was a small bag made of cloth, silk and satin. Wealthy people and people in the court using gold and silver made sachet for perfume storage hung on the bed curtain and on the body. As early as pre-Qin period, there was the custom of wearing sachets. In ancient China, children wore sachets on Dragon Boat Festival to drive away evil and plague. It was used for decoration as well. A woman gave her sachet to a young man as a token of love. Sachets were made of plain gauze among common people. They were embroidered mandarin duck, lotus flower and other auspicious patterns. In Ming and Qing dynasties, sachets prevailed among women and men to express their love, thus sachets became very intimate items which were worn by the young married.

II. Translate the following paragraph into English.

刺绣工艺是中国优秀的传统手工艺之一，至今已有数千年的历史。中国的刺绣源远流长，在世界服饰文化宝库中占有重要的位置，是中华民族智慧的结晶。中国的刺绣工艺几乎遍布全国各地，不同地区的刺绣风格、用线、用色、针法等各具特色，其中比较有代表性的刺绣有京绣、苏绣、湘绣、蜀绣和粤绣。刺绣工艺早在清代就已在旗袍制作上广泛应用，精美的纹样为旗袍锦上添花。

Paragraph Writing

The structure of a narrative essay

In this unit, you will learn how to write a narrative. At times we make a statement clear by relating in detail something that has happened. In the story we tell, we present the details in the order in which they happened. A person might say "I was embarrassed yesterday", and then go on to illustrate the statement with a narrative. His details could make his moment of embarrassment vivid and real for us, and we can understand just why he felt as he did.

A narrative, in some sense, is simply a story that illustrates a point. That point is often about an emotion you felt: anger, jealousy, confusion, thankfulness, loneliness, sadness, terror, or relief. It's not difficult to see that a key word illustrates a point and is usually embedded in a narrative. For example, if the key word is *sad*, it may make you think of the time when you visited your ex-girlfriend/boyfriend and found out she/he was dating someone else. In this case, your narrative should present your feeling of sadness. A good way to bring an event to life is to include some dialogs, which helps make a situation come alive. Besides, time signals and feelings are also important. In brief, when you write a narrative, you may go through several steps:

1. Write down whatever comes to your mind about the experience without worrying about spelling or grammar.
2. Write your thesis statement and make sure it contains the emotion you will focus on. For example, "The first day I saw my wife, my college sweetheart, was <u>one of the happiest occasions in my life</u>."
3. Make up a list of all the details involved in the experience. Then number these details according to the order in which they occurred.
4. Write a rough draft based on the listed details. Use time signals such as *first, then, after, next, while, during,* and *finally* to help connect details as you move from the beginning to the middle and to the end of the narrative. Make sure to include not only what happened but also how you felt about what was going on. You may make the experience vivid by using dialogs.

Now, read the sample essay and see how the story develops with details.

Topic

A heartbreaking story

Introduction

Thesis statement: I felt heartbroken when I found out what had happened on Melisa's side.

Body

Detail 1: When I saw Melisa had a picture taken together with a guy

Detail 2: What I conversed with Melisa regarding the picture

Detail 3: How I felt and how I reacted to the incident

Conclusion

I walked away understanding what was meant by a broken heart.

Sample essay

Melisa and I had gotten engaged in August, just before she graduated from college. One month ago, I drove up to her campus dormitory to see her as a surprise, but the trip turned out miserable. I felt heartbroken when I found out what had happened on her side.

When I knocked on the door of her dorm room, she was indeed surprised, but not in a pleasant way. I asked Melisa how classes were going, and at the same time I tried to pull off my heavy sweater. As I was throwing it over my head, I noticed a large photo on the wall of Melisa and a tall guy laughing together.

It was decorated with paper flowers and a yellow ribbon, and on the ribbon was written "Melisa and Blake". "What's going on?" I asked, standing there stunned and angry. "Who's Blake?" Melisa laughed nervously and said, "What do you want to hear about, my classes or Blake?"

I felt a terrible pain in the pit of my

stomach, and I wanted to rest my head on someone's shoulder and cry, but I did nothing. Angrily, I pulled on my sweater again. My knees felt weak, and I barely had control of my body. I opened the door, and I wanted to slam the door shut. Instead, I managed to close the door quietly.

Anyway, I walked away understanding what was meant by a broken heart. Hope this feeling will never come to me again in the rest of my life.

Write an essay of no less than 150 words on one of the following topics. One topic has on outline that you can follow.

Outline

Topic

Learning English grammar

Introduction

Thesis statement: Now I think it is really fun to learn English grammar.

Body

Detail 1: Adding a comma could show an entire difference in meaning.
Detail 2: A grammar mistake could make a world of difference in meaning.

Conclusion

English grammar is truly a mysterious world of art.

More topics

Topic 1: My favorite restaurant
Topic 2: An unforgettable English lesson/teacher

Origin and Evolution of Qipao

1   China has long **been referred to as** the Kingdom of Dresses. Using their wisdom and talents, Chinese people have created countless dresses and adornments with Chinese characteristics. China has many ethnic groups and different places have their own folk customs, including those in clothing. There is no doubt that clothing culture of China is the result of the **merging** of the costumes of all **ethnic** groups in the country.

2   Qipao **derived** its name **from** the fact that it evolved from the gown-style dress worn by Manchu people known as Qi people in the Qing dynasty. The clothes worn by Chi people were called Qi-clothing. Before 1644, Manchu people lived in the Changbai Mountain and the Songhua River Valley in the northeast of China. As **nomadic** people, they made a living by fishing and hunting. The earliest Qi-clothing had a simple style and structure, with one straight and **cylindrical** piece covering the whole body. It was apparently different from the two-piece design of Han clothing.

3   A typical Qi-clothing at that time had a round collar, a large front piece, narrow sleeves, several buttons and a waist belt. The narrow **cuff** was the most obvious feature, and its shape was like a horse **hoof**, hence its two other names were Horse Hoof Cuff and Arrow Cuff. In winter, the sleeves could cover the back of the wearer's hands, keeping warm but not **hampering** arrow shooting. In normal times, the sleeves could be rolled up as a decoration. Its waist belt could not only help **preserve** the warmth, but allow storage of food in the front piece. In addition, the gown had **slits** on four sides, which is good for horse riding. Men's long gowns were mostly blue, gray or green; and women's, white.

4   With the regime getting **stabilized**, the Qing rulers began **stipulating** and

improving clothing institution of the Qing dynasty by forcing a clothing reform on people. They ordered that all people should wear Qi-clothing **regardless of** their gender and ethnic group.

5   After entering the Central Plains, Manchu people experienced a huge change in their life style. They turned from a nomadic and hunting life style to a stable farming life style. In addition, the non-stop cultural exchange and mutual influence between Han and Manchu peoples also **contributed to** the change of Manchu clothing.

6   At that time, man wore a Ma Gua over the gown, while woman wore qipao, whose cuffs had gradually transformed from tight to loose and were known as "**reversed** large sleeves". The previous heavy and wear-resisting fur was replaced by lighter cotton cloth and silk. The color of qipao also changed from simple to **diversified** and bright. There were simple pattern adornments, and later the lace trimmings and embroidered borders were seen on the collar, cuff, and front pieces of the clothes.

7   Although the 1911 Revolution **toppled** the rule of the Qing dynasty, the female dress **survived** the political change. With **succeeding** improvements, it has become the traditional dress for Chinese women. qipao has undergone great changes in the last century. These changes are not only in form, but more importantly, they represent cultural development and maturity. With the change of people's life style and **aesthetic** interests, qipao has **evolved into** diversified styles.

8   At the end of 19th century and the beginning of 20th century, western clothing was spread to China together with the advanced production technologies. In the first decade of 20th century, life style and dressing concept of Chinese people were changing **drastically**. Fewer people chose to wear qipao, while only noble Manchu families kept on wearing it. Qipao at that time was still developing quietly, although its overall contour was still the loose design with straight lines popular in the late Qing dynasty. Influenced by the social tendency that **esteemed** natural elements, qipao

was simplified. With its sleeves tightened up, **hem** lifted to ankles and adornments greatly reduced, the overall style tended to be simple and elegant.

9   Around 1920, the May Fourth Movement took place in China. Chinese people began accepting some new ideas and concepts, which promoted the **emancipation** of women in mind and body. During that period, qipao began to possess tightened and lowered waist line and more **highlighted** curves on chest, waist and **hips**. Meanwhile, hem became shorter and was lifted to above knees, exposing women's **shanks**. In addition, qipao was made of more diversified materials, such as **flannel**, woolen fabrics, **gauze** and lace. The 1930s and 1940s saw the **full-swing** development of qipao, which finally **gave birth to** a new style of it worn by modern women and **laid the foundation** for qipao to become a classical female dress in Chinese clothing history.

10   In the middle of the 20th century, qipao was simplified and gradually evolved into a fixed style, featuring a standing collar, **contracted** waistline, low slits and a few decorations. By the end of 1950s, qipao had been substituted by Zhongshan suit or Mao suit. By 1960s, it totally disappeared in China.

11   In the early 1980s, qipao rose again in China and became the official dress for women on formal occasions. As qipao can best **showcase** the **stature** and **temperament** of women in the east, it has been treasured by people and loved by women from the mid-1990s again. It is accepted as the formal attire in social gatherings and diplomatic events of China.

12   Since its **advent** in the Qing dynasty, qipao has been developing for three hundred years and it has reached a new height after some **subversive** modifications. Nowadays, with its brand-new presence, qipao shines on all major fashion stages in the world, fully demonstrating its strong vitality and charming **glamour**.

Notes

Qipao, also called Chinese cheongsam, is a straight, tightly fitting silk dress with a high neck and short sleeves and an opening at the bottom on each side, worn by Chinese women.

Zhongshan suit　中山装

New words and phrases

merge /mɜːdʒ/	*v.* ~ **(with/into) sth	~ A with B	~ A and B (together)** to combine or make two or more things combine to form a single thing（使）合并；结合；并入
ethnic /ˈeθnɪk/	*a.* connected with or belonging to a nation, race or people that shares a cultural tradition 民族的；种族的
nomadic /nəʊˈmædɪk/	*a.* roaming about from place to place rather than living in one place all the time 游牧的
cylindrical /səˈlɪndrɪkl/	*a.* shaped like a cylinder 圆柱形的；圆筒状的
cuff /kʌf/	*n.* the end of a coat or shirt sleeve at the wrist 袖口
hoof /huːf/	*n.* (*pl.* hoofs or hooves) the hard part of the foot of some animals, for example horses（马等动物的）蹄
hamper /ˈhæmpə(r)/	*v.* [often passive] to prevent sb from easily doing or achieving sth 妨碍；阻止；阻碍
preserve /prɪˈzɜːv/	*v.* ① to keep a particular quality, feature, etc.; to make sure that sth is kept 保护；维护；保留 ② [often passive] to keep sth in its original state in good condition 维持……的原状；保存；保养
slit /slɪt/	*n.* a long narrow cut or opening 狭长的切口；长而窄的口子；狭缝；裂缝

stabilize /ˈsteɪbəlaɪz/	*v.* to become or to make sth become firm, steady and unlikely to change; to make sth stable （使）稳定；稳固
stipulate /ˈstɪpjuleɪt/	*v.* (*formal*) to state clearly and firmly that sth must be done, or how it must be done 规定；明确要求
reverse /rɪˈvɜːs/	*v.* to change sth completely so that it is the opposite of what it was before 颠倒；彻底转变；使完全相反
diversified /daɪˈvɜːsifaɪd/	*v.* to change or to make sth change so that there is greater variety （使）多样化；变化；不同
topple /ˈtɒpl/	*v.* to overthrow, to make sb lose their position of power or authority 打倒；推翻；颠覆
survive /səˈvaɪv/	*v.* to continue to live or exist despite a dangerous event or time 幸存；幸免于难；艰难度过
succeed /səkˈsiːd/	*v.* to come next after sb/sth and take their/its place or position 接替；继任；随后出现
aesthetic /iːsˈθetɪk; esˈθetɪk/	*a.* concerned with beauty and art and the understanding of beautiful things 审美的；有审美观点的；美学的
drastically /ˈdræstɪkli/	*ad.* extreme in a way that has a sudden, serious or violent effect on sth 极端的；急剧的；严厉的；猛烈的
esteem /ɪˈstiːm/	*v.* (not used in the progressive tenses 不用于进行时) [usually passive] to respect and admire sb/sth very much 尊重；敬重
hem /hem/	*n.* the edge of a piece of cloth that has been folded over and sewn, especially on a piece of clothing （衣服等的）褶边；卷边
emancipation /ɪˌmænsɪˈpeɪʃn/	*v.* [often passive] ~ **sb (from sth)** to free sb, especially from legal, political or social restrictions. 解放；使不受（法律、政治或社会的）束缚

highlight /ˈhaɪlaɪt/   *v.* to emphasize sth, especially so that people give it more attention 突出；强调

hip /hɪp/   *n.* the area at either side of the body between the top of the leg and the waist; the joint at the top of the leg 臀部；髋

shank /ʃæŋk/   *n.* the part of an animal's or a person's leg between the knee and ankle（动物或人的）胫；小腿

flannel /ˈflænl/   *n.* [U] a type of soft light cloth, containing cotton or wool, used for making clothes 法兰绒

gauze /ɡɔːz/   *n.* [U] a type of light transparent cloth, usually made of cotton or silk 薄纱；纱罗（通常由棉或丝织成）

full-swing /fʊlˌswɪŋ/   *a.* reaching the highest level of activity 处于全盛阶段

contract /ˈkɒntrækt/   *v.* to become less or smaller; to make sth become less or smaller（使）收缩；缩小

showcase /ˈʃəʊkeɪs/   *v.* display or present to its best advantage 展示

stature /ˈstætʃə(r)/   *n.* a person's height 身高；个子

temperament /ˈtemprəmənt/   *n.* [C, U] a person's or an animal's nature as shown in the way they behave or react to situations or people（人或动物的）气质；性情；性格；禀性

advent /ˈædvent/   *n.* [sing.] **the ~ of sth/sb** the coming of an important event, person, invention, etc.（重要事件、人物、发明等的）出现；到来

subversive /səbˈvɜːsɪv/   *a.* trying or likely to destroy or damage a government or political system by attacking it secretly or indirectly 颠覆性的；暗中起破坏作用的

glamour /ˈɡlæmə(r)/   *n.* physical beauty that also suggests wealth or success 迷人的美；魅力

be referred to as	be called as 被称为；被称作；被认为是
derive from	stem from, come from 源于
regardless of	paying no attention to sth/sb; treating sth/sb as not being important 不管；不顾；不理会
contribute to	lead to 有助于；促成
evolve into	develop into 发展成；进化成
give birth to	create or produce, have a baby 生产；产生
lay the foundation for	build or put sth for the future use 为……奠定基础
be substituted for	take the place of, be replaced by 代替

■ Reading Comprehension

Understanding the text

Answer the following questions.

1. A typical Qi-clothing worn by Manchu people before 1644 is _____.

    A. a simple style with two-piece design

    B. a simple style with a large piece covering the whole body

    C. a loose style with simple pattern adornments

    D. a diversified and bright style

2. The name of Horse Hoof Cuff derives from the fact that _____.

    A. people have a tradition of naming it

    B. it is shaped like a horse hoof

    C. they are nomadic people

    D. it is an obvious feature of Qi-clothing

3. The traditional Qi-clothing has many advantages for nomadic people, which one is not mentioned in the passage?

    A. It can keep warm.

    B. It can be convenient for arrow shooting and horse riding.

    C. It can store food in the front piece.

    D. Its design is as beautiful as that of Han clothing.

4. Due to cultural exchange and mutual influence between Han and Manchu peoples, the change of Manchu clothing has taken place in the following aspects expect _____.

   A. raw materials

   B. color

   C. shape

   D. function

5. The full-swing development of qipao is in _____.

   A. 1920s

   B. 1930s and 1940s

   C. 1940s and 1950s

   D. 1980s and 1990s

6. Qipao is treasured by women in the mid-1990s, the main reason is _____.

   A. it is a symbol of noble family

   B. it is a classical dress for women in Chinese clothing history

   C. it is a trend in the society

   D. it can best present the beauty and nature of the oriental women

Critical thinking

Work in pairs and discuss the following questions.

1. What is the origin of qipao?
2. Describe briefly the evolution of qipao according to the text.
3. Why is qipao the representative of Chinese clothing?

Language Enhancement

Words in use

Fill in the blanks with the words given below. Change the form when necessary. Each word can be used only once.

merge	drastically	stabilize	dwindle	stipulate
preserve	hamper	nomadic	glamour	highlight

1. It's imaginable that a lack of conversational flow may _____ the integration of immigrants who have not completely mastered the language of their new country yet.

2. In recent years, many of them have been obliged to give up their _____ lifestyle, but they continue to depend mainly on nature for their food and clothes.

3. The report _____ the major problems facing society today.

4. Although her illness is serious, her condition is beginning to _____.

5. The manufacturing economy declined _____ in October for the sixth consecutive month.

6. The factory's workforce has _____ from over 4,000 to a few hundred.

7. Some geologists believe that these people will continue to move westward and eventually _____ with East Asia.

8. Now that she's a flight attendant, foreign travel has lost its _____ for her.

9. The job advertisement _____ that the applicant must have three years' experience.

10. All efforts to _____ the peace have failed.

Expressions in use

Fill in the blanks with the expressions given below. Change the form when necessary. Each expression can be used only once.

derive from	regardless of	give birth to	be substituted by
be referred to	evolve into	lay the foundation	contribute to

1. With the enforcement of encouraging three-child policy in China, some couples are willing to _____ a third child.

2. It's hopeful that each of us can _____ the future of the world.

3. Some of the cultural conflicts _____ a lack of mutual understanding.

4. We have been called pessimistic, but we like to _____ as realistic.

5. With the development of scientific technology, many traditional crafts might _____ the robots in the future.

6. Everyone has the right to good education _____ their religion, color or creed.

7. What we are learning at school will _____ for the future.

8. Based on the effects of online teaching during the pandemic, at least at this stage, it's unlikely to _____ the dominant form of education.

Sentence structure

I. Complete the following sentences by translating the Chinese into English, using "While sth/sb is/does … , sth/sb else is / does …" structure.

| Model: | At that time, men wore a long gown and a Ma Gua over the gown, _____ (而女人身着旗袍) . |
| | →At that time, men wore a long gown and a Ma Gua over the gown, while women wore qipao. |

1. Encouragement will always keep students' minds open to new knowledge, _____ . (而惩罚则使学生充满挫败感)

2. Most digital camera owners are male, _____ _____ (而女性偏爱用手机拍照).

3. _____ (玛丽很擅长人文学科), her younger sister is absolutely hopeless.

II. Rewrite the following sentences by using "regardless of … "

| Model: | They ordered that all people, whatever their gender and ethnic group, should wear Qi-clothing. |

1. The club welcomes all new members no matter how old they are.

   _____

   _____ .

2. No matter the dream is big or small, the goal is high or low, from now on, take prompt actions.

   _____

   _____ .

3. No matter who they are or what they do, people spend more time at work than at any time in the fast-developing society.

   _____

   _____ .

Extensive Reading

The History of Kimono, Traditional Japanese Clothing

1   The Japanese traditional costume is the symbol of the nationality. We can see many kinds of styles, admirable textiles and fabrics in their costume all around the world. The Japanese traditional costume is closely linked to the custom, which has practiced in that area so that the respective dress pattern represents the nation or religion to express who they are. European traditional costume has similar dress pattern and a corset has been used to emphasize women's bodyline to appeal an "hourglass." However, the Japanese kimono conceals women's bodyline. This difference on both costumes shows the different approach how people and society perceive

on dress and women. This is to say that the Japanese traditional costume is not only the dress to decorate on women's body, but also to express the identity of their life and culture.

2   The kimono is the Japanese traditional garment. Originally the word "kimono" referred to all types of clothing, but it has come to mean specifically the full-length traditional garments. Today, kimonos are most often worn by women, and on special occasions. Few older women and even fewer men still wear kimonos on a daily basis. Kimono is worn most often at weddings, tea ceremonies, and other very special or very formal occasions. The name for traditional Japanese clothing is wafuku, ("wa" means Japanese and "fuku" means clothing) and it's yofuku for western or any non-Japanese style. Of all wafuku, the kimono is the most instantly recognized Japanese garment and considered the national costume of Japan. Since kimonos are traditionally made from a single bolt of cloth, larger sizes are difficult to find and very expensive to have made, requiring special looms. Very tall or heavy people, such as sumo wrestlers, have kimono custom-made. Today, both men's and women's kimonos are increasingly available in different sizes. There are many rules when wearing kimono.

3   There are styles of kimono for various occasions, ranging from extremely formal to very casual. The level of formality of women's kimonos is determined mostly by the pattern and fabric, and also the color. Young women's kimonos have longer sleeves and tend to be more elaborate than similarly formal older women's kimono. Men's kimonos are usually one basic shape and are mainly worn in subdued colors. Formality is also determined by the type and color of accessories, the fabric, and the number or absence of kamon (family crests). Silk is the most desirable, and most formal, fabric; cotton is more casual. Kimonos made of modern fabrics such as polyester are also more casual than silk. Choosing a kimono by its formality for special occasions is very important.

4   Most of the today's Japanese people live in western style clothing. But traditional Japanese clothing is treated as a dress at an important event including a wedding ceremony, and young people who wear yukata in summer increase recently. The prototype of the current kimono can be found in Heian Period of about 1,200 years ago. Then people have used various fabrics, patterns, and colors for the clothing. Kimono used advanced technology and materials became an area of Japanese art. Let's go back to the ancient times before the establishment of the kimono and look at the history of Japanese clothing.

5   By the time of Jōmon Period (c.2500–1500BCE) when hunting living, and Yayoi Period (c.300BCE–c.250CE) when the rice crop started, the Japanese had still worn the simple and coarse clothes without any pattern. In Jōmon Period, animals' fur was used as a material for clothing at the beginning. But hemp gradually became the leading role, and people with high status were wearing silk clothes in Yayoi Period.

6   During Tumulus Period (250–552CE), the interaction with the Korean Peninsula and China increased and the shape became affected by the cultures. The women tightened the band at a position higher than the waist and put on clothes like a long skirt. The men wore trousers tied with a string under the knee.

7   In Asuka Period (552–645CE) and Nara Period (710–784CE), Japan adopted the political system and the culture from China. Therefore, the clothing changed to those which were very close to the country. The most characteristic feature of women's clothes at this time was the skirt. The folds were attached to the hem of the skirt, and the form spread as it goes down than before. Men generally wear hats, and in daily life, they were wearing a simple dress. But members of the nobility and the imperial court wore the round-necked robe and the men's skirt. The robe's collar was based on their classes.

8   Japanese traditional kimono appeared in Heian Period. It was because the

government banned trade and diplomacy with China in this era. It was the aristocracy who became the center of politics and culture, they liked clothes emphasized on gorgeous appearance rather than functionality. The aristocratic women wore a coat called "oosode" (大袖 , big sleeves) at the top and "kosode" (小袖 , small sleeves) in layers with different colors under it. The kosode is the origin of the present kimono. The common women wore only the kosode. Since they wore twelve kimonos, the formal dress is called "Juni-hitoe"(12 layer dress). The total weight of them was about 10kg. The dress for male aristocrats were lighter than females and many were in moderate color.

9   In the Heian Period, the clothing was given the top priority to elegant appearance. However, since the warrior and their families were more active than nobles, the kimono became simpler to move easily. Kosode, which played the role of underwear for 12 layer dress, became independent and its number gradually decreased. As the layers decreased, the women had to tighten the collars firmly and they started to use "Obi"(band). Although the existence of Obi had been ambiguous in Heian Period it came out clearly to the front around 1300s or 1400s. In the 1500s, the features of the present kimono were almost established, and the technique of weaving and dyeing developed greatly. On the other side, most of the samurais wore kataginu (short sleeveless garment made of hemp) and hakama (pleated and divided skirt made in fine strips) until Edo Period.

10   In the Edo period (1603–1867) when Japan had little diplomacy and exchange with other countries, the people who led the culture were town people. They began to use a band that was wider and more having a presence than before and it became almost the same style as the current kimono. In addition, the golden threads and the gorgeous patterns were given to the clothing.

11   Western culture flowed into Japan all at once after the opening to the countries in 1854, and people gradually changed their clothing to western style from the upper class. However, many people wore the clothes

combined the characteristics of the West and Japan until the middle of Showa Period. After World War II, most Japanese began wearing the western clothes in daily life, and kimonos only on the days of the event and special occasions.

Unit 5

Fashion Marketing

The art of marketing is largely the art of brand building.

—*Philip Kotler*

The real task of marketing is to make sales promotion redundant.

—*Peter F. Drucker*

Pre-Reading Activities

1. Listen to a short talk about fashion marketing and fill in the blanks based on what you hear.

Fashion marketing is defined as the process of planning and (1) _____ goods or services in the fashion industry to the right (2) _____ at the right time and place with the use of marketing and advertising tools and (3)_____. Like other branches of marketing, fashion marketing involves the following marketing activities: Identifying your target customer base and understanding and anticipating their specific fashion needs; predicting, tracking, and (4) _____ trends in styles, colors, sizes, fabrics, package designing; forecasting sales in the current fashion trend cycle; conceptualizing, (5) _____, and developing fashion products; determining resources, and selecting and buying the right (6) _____; planning the pricing, promotion, and distribution of products; crafting advertising campaigns and promotional activities to (7) _____ buyers; creating appealing and exciting fashion shopping experiences. Fashion marketing is a dynamic industry as the fashion cycle keeps evolving. Therefore, fashion marketers need to make quick (8) _____ and adapt to fast-changing trends to be successful in this profile.

2. Discuss the following questions with your partner.

(1) Do you have any experience of marketing? If yes, how do you feel?

(2) What is fashion marketing and why do we really need it?

(3) Why is visual marketing so important? How can people create effective visual marketing for the business?

How Can Fashion Brands Smartly Use Visual Marketing?

1   People say that "a photo is worth a thousand words." Well, for fashion brands, it is worth more than that. Words aren't enough to attract new customers and engage the existing customers these days. For the last decade or so, every brand has made visual content a primary part of their marketing strategy. It is evident from their website, blogs, social media pages, and almost every digital channel.

2   You might be wondering why the popularity of visual content surged in the digital age. The answer to it lies in science. The research into human behavior has confirmed that our brain can process visual content 60,000 times faster than text. There is little left to defy that "seeing is believing."

3   Although text-based content has remained a key component of marketing, in 2020, visual content is the king. Since many brands produce content daily, it is the visual content that will enable you to cut through the clutter and make yourself stand out.

4   We will share the tactics that fashion brands can be used to take their marketing campaign to the next level. But before that, let's discuss the importance of visual content in marketing.

5   Visual content marketing refers to the type of marketing that relies on visual assets like images, videos, and infographics. Research has proven that 90 percent of the information that our brains receive is visual. So, if you aren't using visual content in your marketing strategy, you are pushing most of the audience away. Visual content is particularly noteworthy on social media, where it outperforms plain text by a wide margin. As the study of the top 100 brands by CoSchedule found that more than

80 percent of all content they published on Facebook were photos and videos. Additionally, visual content accounted for over 90 percent of all interactions and social media engagements.

6   **In essence**, visual content marketing isn't much different from text-based content marketing. You need to **come up with** a strategy first and **focus on** reaching the right people at the right time with the right message. Similar to text-based content, the context in which images are used will play a vital role in the campaign. Different **platforms** have different visual content requirements, and different **demographics** prefer different visual content. Therefore, you need to have **a broad range of** visual content, from infographics to **GIFs** and engaging videos, to meet these requirements. It is also important to note that visual content requires more time and resources than written content. You also need to be patient and wait for a while before it starts **generating** a **decent** return on investment. However, you need to analyze and **monitor** the key **metrics** at the same time, to ensure that you are heading in the right direction. Below are a few ways you can use visual content to market your fashion brand.

7   It is not a bad idea for fashion brands to put together a style guide that tells the audience about the everyday use of their items. This will give the visitors an idea of how to wear your fashion **items** and **incorporate** them **into** their wardrobe. You can create separate style guides for seasonal, wedding, party, or holiday themes. They can be listed as a separate section on your website where you provide detailed information about each item in the collection. Use catchy images to attract the interest of the customers. For example, Leather Skin Shop, an online leather jacket store has increased user engagement on its social media posts by frequently posting style guides, fashion blogs and other tips to help their customers.

8   User-generated content refers to the type of content that is created by unpaid contributors. It can include photos, videos, tweets, **testimonials**, and blog posts that promote a brand. It is the most effective form of marketing because consumers create free content and share it **on behalf**

of the brand. However, a brand needs to retweet the content on its social media page or website to increase efficacy. Many brands use user-generated content, but the one that excels at it is GoPro. It rarely uses its logo on visual content because the images themselves yell "GoPro." They regularly share stunning visuals taken by their customers, which helps them in branding the company.

9   Data can be boring, but there are appealing ways to present it. Infographics are one of them. Due to their visual appeal, infographics are loved by users. In fact, they are shared three times more than any other type of visual content. Fashion brands can utilize them to communicate information quickly and clearly. Since infographics are eye-catching, they can significantly boost your shares and maximize your sales. Some of them seem to be complicated, but with the right tools, you will realize that they aren't difficult at all. Resources like Piktochart and Canva offer a quick and simple way to present all your data in the form of infographics.

10   Presentations are no longer used for business purposes only. Today, they are used as a marketing tool by the brands. You can use platforms like SlideShare to expand your outreach through compelling presentations. If a topic is too lengthy to fit in an infographic, you can craft a presentation and present your data to your audience. When crafting a presentation, make sure to remain consistent with your design, colors, and fonts. Perform spelling and grammar checks and keep the sources as credible as possible.

11   You do not need to be a graphic designer to create visually appealing content. Use the right tools and produce quality content that resonates with the audience. However, don't put out content just for the sake of a content calendar. Understand the need of the audience and publish content that will be appreciated.

Notes

CoSchedule is a social media calendar tool that makes it easier to plan and execute social media marketing strategies. In August 2013, CoSchedule opened its official office in Bismarck, North Dakota. You can simplify your planning process by building templates for specific actions in the tool and then adding them to your calendar. In August 2016, CoSchedule was officially launched in more than 100 countries.

GIF （Graphic Interchange Format）可交换的图像文件

GoPro 美国运动相机厂商

PiktoChart 马来西亚一款信息图设计和展示工具

Canva 澳大利亚一款支持多平台在线平面设计工具

SlideShare 一个存储与展示幻灯片的网站

New words and phrases

engage /ɪnˈɡeɪdʒ/	v. ① occupy, attract, or involve (someone's interest or attention) 引起（某人的兴趣或注意）；吸引 ② (~ **sb in**) cause sb to become involved in (a conversation or discussion) 使某人加入（谈话，讨论） ③ (~ **in**) participate or become involved in 参加，从事；卷入
primary /ˈpraɪməri/	a. ① of chief importance; principal 首要的；主要的 ② earliest in time or order of development 最初的，原始的
strategy /ˈstrætədʒi/	n. ① [C] a plan of action or policy designed to achieve a major or overall aim 战略 ② [U] the process of planning sth or putting a plan into operation in a skillful way 策划，部署
evident /ˈevɪdənt/	a. plain or obvious; clearly seen or understood 明显的；显而易见的
surge /sɜːdʒ/	v. ① (of a crowd or a natural force) move suddenly and powerfully forward or upward（人群、自然力）汹涌；奔腾 ② increase suddenly and powerfully 急剧上升；激增

confirm /kən'fɜːm/   *v.* ① establish the truth or correctness of (sth previously believed, suspected, or feared to be the case) 证实，确定（相信、怀疑、担心的事）② state with assurance that a report or fact is true 证实（报道，事实）

defy /dɪ'faɪ/   *v.* ① openly resist or refuse to obey 公然反抗；藐视 ② appear to be challenging (sb) to do or prove sth 向（某人）挑衅（或挑战）

component /kəm'pəʊnənt/   *n.* a part or element of a larger whole, especially a part of a machine or vehicle（尤指机械或车辆的）部件，零件；（构成整体的）组成部分

clutter /'klʌtə/   *n.*[U] a collection of things lying about in an untidy mass 杂物 [C] an untidy state 杂乱

v. crowd sth untidily 使乱成一团；杂乱地堆满

tactic /'tæktɪk/   *n.* an action or strategy carefully planned to achieve a specific end 策略；招数

campaign /kæm'peɪn/   *n.* ① a series of military operations intended to achieve a particular objective, confined to a particular area, or involving a specified type of fighting 战役；战斗 ② an organized course of action to achieve a particular goal 运动（为取得某目的而组织的活动）

asset /'æset /   *n.* ① a useful or valuable thing, person, or quality（有用的或宝贵的）物品；人才；品质；优点 ② (*usually pl.*) property owned by a person or company, regarded as having value and available to meet debts, commitments, or legacies （个人或公司的）资产（可以用于偿付债务、承付款项或遗赠）

noteworthy /'nəʊtwɜːðɪ/   *a.* interesting, significant, or unusual 令人关注的，值得注意的；重要的；不寻常的

outperform /aʊtpə'fɔːm/   *v.* ① perform better than 做得比······好 ② (of an investment) be more profitable than（投资）比······回报更高

platform /'plætfɔːm/   *n.* ① a raised level surface on which people or things can stand 台，平台 ② the type of computer system or the software that is used 计算机平台

demographic /ˌdemə'ɡræfɪk/   *a.* relating to the structure of populations 人口结构的；人口统计的

   n. [*pl.*] statistical data relating to the population and particular groups within it 人口统计数据

generate /'dʒenəreɪt/   *v.* ① cause (sth, especially an emotion or situation) to arise or come about 引发（某种情感）；造成（某种情况）；使······发生 ② produce (energy, especially electricity) 产生（能量，尤指电）

decent /'diːs(ə)nt/   *a.* not likely to shock or embarrass others 得体的；恰当的

monitor /'mɒnɪtə/   *v.* ① observe and check the progress or quality of sth over a period of time; keep under systematic review 检测；监测 ② maintain regular surveillance over 监视；监控

metric /'metrɪk/   *n.* a system or standard of measurement 计量体系；衡量标准

item /'aɪtəm/   *n.* ① [C] an individual article or unit, especially one that is part of a list, collection, or set 项目；条款 ② a single article or object 一件商品（或物品）③ a piece of news or information 一条新闻（或消息）

testimonial /ˌtestɪ'məʊnɪəl/   *n.* [C] a formal statement testifying to sb's character and qualifications（品德、资格的）证明书；介绍信；推荐信

efficacy /'efɪkəsɪ/   *n.* [U] the ability to produce a desired or intended result 功效；效力；效能

yell /jel/   *v.* give a loud, sharp cry 叫喊；叫嚷；号叫

stunning /'stʌnɪŋ/   *a.* ① extremely impressive or attractive 绝妙的；很吸引人的 ② extremely surprising or shocking 令人惊奇万分的

appealing /ə'piːlɪŋ/   *a.* attractive or interesting 吸引人的；有趣的

present /'prez(ə)nt/   *v.* ① **~ sb with sth/~ sth (to sb)** give sth to sb formally or ceremonially 郑重赠予；授予；颁发 ② show or describe sth/sb in a particular way (以某种方式) 展现；显示；表现

utilize /'juːtəlaɪz/   *v.* make practical and effective use of 利用

significantly /sɪg'nɪfɪk(ə)ntli/   *ad.* in a way that is large or important enough to have an effect on sth. or to be noticed 有重大意义地；显著地

boost /buːst/   *v.* help or encourage sth to increase or improve 增长；推进

maximize /'mæksɪmaɪz/   *v.* make as large or great as possible 使……达到最大（最高），使……最大化

complicated /'kɒmplɪkeɪtɪd/   *a.* ① consisting of many interconnecting parts or elements; intricate 结构（或成分）复杂的；费解的 ② involving many different and confusing aspects 错综复杂的；混乱的，令人困惑的

presentation /prez(ə)n'teɪʃ(ə)n/   *n.* ① a demonstration or display of a product or idea（产品，观点）展示；表述；描述 ② an exhibition or theatrical performance 展览；上演；演出；表演

outreach /'aʊtriːtʃ/   *v.* reach further than 超出……的限度；超过……的范围；超越；胜过

     n. the extent or length of reaching out 能达到的范围

compelling /kəmˈpelɪŋ/   *a.* ① that makes you pay attention to it because it is so interesting and exciting 激发兴趣的；引人入胜的 ② not able to be refuted; inspiring conviction 无法驳倒的；令人信服的 ③ not able to be resisted; overwhelming 无法抵抗的；势不可挡的；压倒性的

lengthy /ˈleŋθi/   *a.* (especially in reference to time) of considerable or unusual length, especially so as to be tedious（尤指时间）过长的；冗长的

craft /krɑːft/   *n.* ① [C, U] an activity involving a special skill in making things by hand 工艺；手艺 [sing.] the skills needed for one's work 手艺；技能；技巧

v. exercise skill in making sth 精心制作

consistent /kənˈsɪstənt/   *a.* ① compatible or in agreement with sth 一致的；符合的；和谐的；协调的 ② (of an argument or set of ideas) not containing any logical contradictions（论点，观点）不矛盾的；连贯的；一致的

font /fɒnt/   *n.* a set of type of one particular face and size（字体、字号相同的）一副铅字；字型

credible /ˈkredəbl/   *a.* able to be believed; convincing 可信的；令人信服的

graphic /ˈɡræfɪk/   *n.* a graphical item displayed on a screen or stored as data（屏幕上显示或作为数据储存的）图表（或图形、图案等）

resonate /ˈrezəneɪt/   *v.* ① produce or be filled with a deep, full, reverberating sound 回响；回荡 ② evoke or suggest images, memories, and emotions 引发（想象）；唤起（回忆）或引起（情感共鸣）

appreciate /əˈpriːʃieɪt/   *v.* ① recognize the full worth of 认识到……的全部价值；赏识；重视；欣赏 ② be grateful for sth 为……表示感谢；感激

or so	(of quantities) imprecise but fairly close to correct 大约；左右
lie in	originate (in) 在于……
cut through	travel across or pass over 穿梭；越过
stand out	① be highly noticeable 引人注目 ② be clearly better or more significant than someone or sth 杰出；更突出；更重要
rely on	be dependent on, as for support or maintenance 依靠；依赖
in essence	in nature; at heart 本质上；其实
come up with	think of an idea, plan, reply, etc. 提出；想出
focus on	pay special attention to a particular person or thing instead of others 专注于（特定的人或物）
a broad range of	involving large numbers or a large area; extensive 广泛的
incorporate into	make into a whole or make part of a whole 纳入；收入
on behalf of	in the person of 代表；为了
excel at	be very good at; shine at 突出；擅长于
fit in	go together; adapt to something 适应；适合
resonate with	meet with sb's agreement 与……产生共鸣；达成一致
put out	prepare an issue for public distribution or sale 发行；发布
for the sake of	in the cause of; the purpose of achieving or obtaining 为了

Reading Comprehension

Understanding the text

Answer the following questions.

1. How do you understand the statement: "A photo is worth a thousand words" ?

2. Why has the popularity of visual content soared in the digital age?

3. What did the study of the top 100 brands by CoSchedule find?

4. What is the importance of the visual content marketing?

5. What effective ways are recommended to market fashion brand with visual content?

6. How can you attract customers and make them wear your fashion items?

7. Why is user-generated content the most effective form of marketing?

8. What are the advantages of infographics?

9. What problems should be paid attention to when making a presentation?

10. According to the last paragraph, how can people create visually appealing content for fashion brands?

Critical thinking

I. Presentation.

Collect information about the successful fashion marketing strategies by surfing the Internet after class. The information may include the types and features of the marketing strategies. You are supposed to collect as much information about the strategies as possible. Work in groups and summarize what you have discussed in a report. Then choose a representative from your group to present the report to the class.

II. Work in pairs and discuss the following questions.

1. What kind of visual marketing strategy for fashion brands do you prefer? Why?

2. What do you think of visual content marketing for fashion brands?

Language Enhancement

Words in use

Fill in the blanks with the words given below. Change the form when necessary. Each word can be used only once.

primary	evident	surge	confirm	decent
appealing	boost	consistent	noteworthy	appreciate

1. Mr. White's assistant telephoned to _____ his appointment with the chairman.

2. The local government has taken a range of measures to _____ tourism to stimulate the economy.

3. The idea of having enough money to retire at fifty is very _____.

4. I'd _____ any information you could give me.

5. She ate the biscuits with _____ enjoyment.

6. The course of action which he is following should be _____ with his sense of responsibility as a president.

7. Science has recently made _____ progress especially in this field.

8. The government's _____ aim is to see significant reductions in unemployment.

9. Today, the president wore a _____ high-necked dress to attend the ribbon-cutting ceremony of the company.

10. The shares in the hotel corporation _____ to a record high in the first quarter of this year.

Banked cloze

Fill in the blanks by selecting suitable words from the word bank. You may not use any of the words more than once.

A. advertised	F. effectively	K. evidently
B. positive	G. recognize	L. purchased
C. vital	H. showcase	M. recommending
D. popular	I. strategy	N. resonate
E. potential	J. consuming	O. perception

That being said, digital marketing should still be considered a key part of your overall 1. _____. With so many brands now going digital, it's 2. _____ that you make yours stand out from the crowd. Some ways to achieve this include:

Retarget Visitors: You can target visitors through displaying reminder ads that make them recall a product they saw previously on your site. This can even include lucrative discounts as well as retargeting those who have an item in their cart but haven't 3. _____ yet. These ads help to increase turnover and an effective retargeting campaign can produce a great ROI.

Connect With Influencers: To focus on your target market 4. _____, work with influencers such as fashion bloggers and vloggers who have a following that is similar to the type of customers you want. Influencers are viewed as a trusted voice by their followers and many will prefer this format to more traditional methods of marketing such as television. They can help increase your sales by reviewing or 5. _____ your products while it allows you to have your products 6. _____ to a different group of 7. _____ customers. If you've chosen the right influencer, it can lead to new customers as well as create a repeatable form of digital marketing for fashion brands.

Engage With Your Audience: Build brand loyalty begins with engagement. The key is to constantly engage with your target audience to help improve their 8. _____ of your brand. This enhanced sense of community is what helps drive further sales. The best way to engage effectively is to 9. _____ campaigns that connect and 10. _____ with your audience. This could be through running contests on Facebook, responding quickly to messages on Twitter or promoting giveaways on Instagram that highlight the values and products you have to customers.

Expressions in use

Fill in the blanks with the expressions given below. Change the form when necessary. Each expression can be used only once.

stand out	a broad range of	come up with	incorporate into
excel at	resonate with	on behalf of	for the sake of

1. According to analysis, we are supposed to _____ proposals to improve the effectiveness of the regional investment.
2. They are ready to support public engagement by scientists, and to _____ scientific knowledge _____ the public communications.
3. The song is well-suited to the current economic climate and will likely _____ many listeners.
4. Lily has _____ knowledge because she likes reading extensively.
5. He determined to make great effort to learn English well _____ going abroad.
6. Please allow me to thank you for your offer of help _____ my colleagues.
7. She has some characteristics that make her _____ from the crowd.
8. People rarely _____ tasks they don't enjoy doing or feel passionate about.

■ Translation

I. Translate the following paragraph into Chinese.

Fashion marketing is the process of managing the flow of merchandise from the initial selection of designs to be produced to the presentation of products to retail customers, with the goal of maximizing a company's sales and profitability.

Successful fashion marketing depends on understanding consumer desire and responding with appropriate products. Marketers use sales tracking data, attention to media coverage, focus groups, and other means of ascertaining consumer preferences to provide feedback to designers and manufacturers about the type and quantity of goods to be produced. Marketers are thus responsible for identifying and defining a fashion producer's target customers and for responding to the preferences of those customers.

II. Translate the following paragraph into English.

营销已经是一个时代的故事：从大众营销经直销时代，到数字营销，再到现在的数据驱动营销。随着营销策略的变化和营销资源的转移，消费者和客户忠诚度也在变化。如今，消费者会使用更多的设备、广告拦截器和其他隐私保护工具，而且仍然有越来越复杂的客户体验期望。别搞错了：如果你没有达到这些期望，客户会找到愿意的人，就像弗雷斯特分析公司 (Forrester Analytics) 的数据那样，54% 的人说他们愿意在任何市场中尝试。

Paragraph Writing

How to develop an advantage/disadvantage essay

Generally speaking, the outline of writing an advantage/disadvantage essay falls in one of these three formats: 1) advantages and disadvantages; 2) advantages only; 3) disadvantages only. When you write such an essay, remember that you are giving information, and a method that is called exposition or expository writing.

Begin your essay by introducing your topic and explaining that you are exploring the advantages or disadvantages of the topic. Mention in your thesis statement the advantages or disadvantages you will discuss in the essay. Use transitions to make your ideas flow smoothly. The following transitional words are very common when you write an advantage/disadvantage essay:

Close your essay with the summary of advantages and/or disadvantages. Even though you are not persuading your readers when you write this type of writing, you are required to add your opinion at the end as your final comment on the topic. For example, if you're talking about the advantages of taking courses of the humanities, you may say in your conclusion:

To sum up, by taking courses in the humanities, students gain more analytical ability, acquire more critical thinking skills, and deal with things more rationally. Students should be aware of the significance of taking these courses.

You can always choose a position when you write about advantage/disadvantage essays. For instance, you don't have to talk about both advantages and disadvantages of taking natural science courses. Instead, you can only focus on one, the advantages of taking these courses or the disadvantages of doing that.

From what is mentioned above, we can see that writing an essay about advantages or disadvantages of an issue requires supporting points. If your essay focuses on advantages of studying the humanities, then several advantages should be provided to support your point.

Structured writing

Read the sample essay and see how the advantage/disadvantage essay is developed.

Topic

The advantages of taking natural science courses

Introduction

Thesis statement: Taking natural science courses benefits students in several ways but two stand out.

Sample essay

At our school, all students are required to take a minimum of six courses in natural sciences. Students majoring in the humanities often have to struggle to get through these demanding courses, so it has been suggested that the requirements be modified. In my opinion, taking natural science courses benefits students in several ways and two stand out.

Body

Advantage 1: Be involved in more active learning.

First, students will be involved in more active learning if they take natural science courses. Obviously, learning natural sciences enables students to question and solve problems without expecting someone to give them specific answers. In this way, they develop their confidence in their own ability. Also, figuring out the process rather than just following directions results in a solution unique to the learners, which will surely help them to go further.

Advantage 2: Be more motivated to learn.

Another important advantage is that students taking natural science courses can be more motivated. That is because natural science courses are more meaningful and practical. Taking natural science courses not only prepares students for their future career but also enables them to make use of their own personal concepts as a basis for understanding. Thus, they are more anxious to grasp the newly discovered ideas.

Conclusion

Being more active and more motivated are the two major advantages for students who take natural science courses.

In brief, being more active and more motivated are the two major advantages for students who take natural science courses. Moreover, by connecting the new information to the real world, students can acquire the skills that are highly valued by their future employers.

Text B

Ten Most Effective Marketing Strategies for Fashion Brand

1   Every fashion brand needs a good marketing strategy in place to help it stay ahead of its competition. A good strategy will not only help them stay afloat in the treacherous sea that is e-commerce, but also grow as others flounder in confusion. As for you, it doesn't matter if you're brand new

to the fashion world or a seasoned veteran—no matter where you lie, you need an effective marketing strategy. Digital marketing for fashion brands specifically has changed dramatically over the past few years and it's vital that your business is constantly looking to adapt and evolve to stay ahead of the competition.

2　Digital marketing for fashion brands is full of a variety of different approaches that your business can take, making it incredibly difficult to know just where exactly you need to start. Here are ten tips to help you create a successful fashion marketing strategy for clothing businesses and brands.

Get a Website

3　*Women's Wear Daily* reported that more than 1,875 fashion stores closed last year, yet according to Statista, revenue in the fashion segment is expected to increase from £360.7 billion in 2018 to £534.5 billion in 2022. This seems contradictory—how is revenue increasing if so many stores are closing? The Internet. Shopping habits have changed and consumers are increasingly preferring to shop online rather than visiting a physical location. For you, this means getting a website if you don't already have one; this means going from a fashion brand to a fashion e-commerce brand.

Have a Mobile Presence

4　Now that you hopefully have a website, we move to digital marketing for the fashion industry. Similar to our propensity for online shopping, we're substituting keyboards for screens and are opting to browse the web with our smartphones rather than desktops. As a fashion brand, this means getting a responsive website that resizes content according to the screen it resides in. Not doing so will lead to subpar experience for your mobile users, which will deter them from further interactions with your brand. This is especially important in the fashion industry because many consumers see things they like as they go about on their day, and if they happen to see something you offer but can't load it on your site, you can find they'll go to a competitor.

Join the Social World

5   Social media marketing for fashion brands is an absolute must. Not only can you use it to connect with your audience and build a relationship with them, but you can also use certain platforms like Instagram to show off and promote your products. Additionally, you can integrate your store with certain channels like Facebook, Instagram and Pinterest, meaning that if your audience sees something they like as they browse through your catalogue, they can buy it right then and there.

Attach an Influencer to Your Brand

6   Speaking of social media, some of the best fashion social media campaigns revolve around influencers. For those who don't know, influencers are people within your industry who have large fan followings and command attention—when an influencer posts, others follow. For example, one of the best ways to promote your clothing line is to get an influencer to tweet pictures of them wearing some of your clothes. At the end of the day, influencer marketing is like high school; influencers are the popular kids and their followers are the other kids who emulate them and want to be just like them.

Leverage the Latest Technologies

7   We're currently in the midst of a technological revolution marked by rapid advancements in technologies like artificial intelligence, such as chatbots. Chatbots act like 24/7 customer service representatives that guide your customers through their journey to checkout as soon as they land on your site. If your users have a question, they're there to answer it. And if they don't have the capability to answer that question, a human rep can take over.

Personalize Everything

8   Shopify reports that 43% of purchases are influenced by personalized recommendations or promotions, and that 75% of consumers want brands

to **personalize** messages, offers and experiences. Apart from the fact that personalization will help you sell more, what's important to note here is that consumers don't just want personalized messages with their name; they want their whole experience with you to be personal. They want you to know them and their **preferences**, and if you're going to promote something, you'd better make sure it's something they're interested in. When it comes down to it, personalization is what creates successful fashion marketing campaigns.

Remarket to Your Users

9   The beauty of remarketing is that you're targeting people who have already visited your website in the past, so right away you know that there's something there for you to work with. Additionally, you're increasing the chances of a **conversion** over a normal ad because you're showing them something they were already looking at on your website, not a **random** item they may or may not be interested in. Remarketing, then, gives you more brand exposure, better audience targeting, higher conversion rates and improved ROI.

Focus on Your Visuals

10   Visuals are important for all brands, and more essential for fashion brands that depend on aesthetics to attract and keep customers. If you upload something that's meant to be seen, whether it's an image or a video, make sure it's of high quality and looks good. These may seem like no-brainers, but you'd be surprised at the amount of brands with **blurry** images that paint their products in an **unflattering** light. Don't be like them—always **capture** as many angles as possible and let your products speak for themselves.

Start a Blog

11   Ask any digital **agency** to name you a few fashion brand must-dos and they will **undoubtedly** tell you to start a blog. As a **hallmark** of content marketing that does everything from improving your organic search

rankings to exposing new users to your brand, blogging is one of the most important marketing tools you have **at your disposal**. For starters, every blog post creates a new landing page for your audience to find you with. Moreover, you can use a blog to **disseminate** any news or updates you may have, introduce new styles with a couple of HD images to show them off, or help your audience with some style guides as the seasons change.

Have a Contest or Giveaway

12　　The last fashion brand marketing strategy we'll cover today is to start a contest or **giveaway**. Whether it's to promote a new product, celebrate a holiday, or simply just because contests and giveaways are excellent ways of acquiring new customers and engaging with existing ones. They're perfect for **fostering** brand **awareness** and growing both your audience and business. **Apart from** the buzz you'll generate around your brand and products, you can have **participants** enter the promotion by **signing up for** an email newsletter, liking or sharing a social page, and so on, which can give you valuable consumer data you can use to tailor your marketing strategy down the line.

13　　There you have it, the ten most effective marketing strategies for fashion brands. Follow one or all (preferably all), they will **definitely** help you out as you either start out or grow in the fashion world. If you're in fashion e-commerce, **get in touch with** a fashion marketing agency to talk about your project!

Notes

Statista is a leading global integrated database, which provides data that involves major countries and economies in the world. Statista's huge data content and powerful search technology can help students, scientific researchers, market research planners and other researchers to find the required statistics and market information in a timely and effective manner, and provide authoritative and effective data for quantitative research. Statista provides users with direct and effective search and data source location, which is an important way for scientific research and survey.

Women's Wear Daily   《女装日报》，美国著名时装行业报

Statista   一个全球综合数据资料库

Pinterest   马来西亚一款信息图设计和展示工具

Shopify   加拿大一家独立的电商网站，成立于 2004 年

HD   (High Definition) 高清晰度

New words and phrases

afloat /əˈfləʊt/	*a.* ① floating on water; not sinking 漂浮在水上的 ② having enough money to pay debts; able to survive 有偿债能力的；能维持下去的
treacherous /ˈtretʃərəs/	*a.* ① hazardous because of presenting hidden or unpredictable dangers 危险的；变化莫测的 ② guilty of or involving betrayal or deception 不可靠的；背叛的；奸诈的
flounder /ˈflaʊndə/	*v.* struggle or stagger helplessly or clumsily in mud or water 无助（或笨拙）地挣扎；蹒跚
confusion /kənˈfjuːʒn/	*n.* ① [U, C] uncertainty about what is happening, intended, or required 不确定；疑惑 ② [U] the state of being bewildered or unclear in one's mind about sth 困惑；糊涂；迷茫
veteran /ˈvetərən/	*n.* a person who has had long experience in a particular field, especially military service 老手（尤指老兵、老战士）
constantly /ˈkɒnst(ə)ntlɪ/	*ad.* all the time; repeatedly 始终；一直；重复不断地

adapt /ə'dæpt/   v. ① make sth suitable for a new use or purpose; modify 使适应；修改；调整 ② alter (a text) to make it suitable for filming, broadcasting, or the stage 改编（文本）◇ ~ **to** become adjusted to new conditions 适应

approach /ə'prəʊtʃ/   v. ① come near or nearer to sb or sth in distance 接近；靠近 ② start to deal with (sth) in a certain way 着手处理

n. a way of dealing with sth 处理事情的方法

revenue /'revənjuː/   n. [U, C] income, especially when of a company or organization and of a substantial nature （尤指公司或组织数额庞大的）收入；收益

segment /'segmənt/   n. each of the parts into which sth is or may be divided 部分；片段

v. divide sth into separate parts or sections 分割；划分

contradictory /ˌkɑːntrə'dɪktəri/   a. mutually opposed or inconsistent 互相对立的；互相矛盾的

propensity /prə'pensɪti/   n. an inclination or natural tendency to behave in a particular way 倾向；习性

substitute /'sʌbstɪtjuːt/   n. a person or thing acting or serving in place of another 代替者；代替物

v. ~ **sb/sth for sb/sth** replace (sb or sth) with another 用（另外的人或物）替代

opt /ɒpt/   v. make a choice from a range of possibilities 选择；做出抉择

browse /braʊz/   v. ① scan through a book or magazine superficially to gain an impression of the contents 浏览（书刊或杂志）；随意翻阅 ② read or survey (data files), typically via a network （尤指通过网络）浏览（数据文件）

reside /rɪ'zaɪd/	*v.* have one's permanent home in a particular place 居住
subpar /ˌsʌb'pɑː/	*a.* below a level of quality that is usual or expected 不到一般（或预期）水平的；低于标准的
deter /dɪ'tɜː/	*v.* discourage sb from doing sth 吓住；制止
competitor /kəm'petɪtə/	*n.* ① an organization or country that is engaged in commercial or economic competition with others（商业或经济上的）竞争对手；竞争者 ② a person who takes part in a sporting contest（体育竞赛）参赛者；比赛者
catalogue /'kætəlɒg/	*n.* a complete list of items, typically one in alphabetical or other systematic order, in particular（尤指按字母或其他系统顺序编排的）目录
	v. make a systematic list of (items of the same type) 将（同类物品）编成目录
command /kə'mɑːnd/	① give an authoritative or peremptory order 命令 ② be in a strong enough position to have or secure 掌握；拥有；可以获得
emulate /'emjʊleɪt/	*v.* match or surpass (a person or achievement), typically by imitation （尤指通过模仿而）赶上；超过
currently /'kʌrəntlɪ/	*ad.* at the present time 目前；时下
advancement /əd'vɑːnsmənt/	*n.* ① [U, C] the process of helping sth to make progress or succeed 促进；推动；发展；进步 ② [U] progress in a job, social class, etc. 提升；晋升

capability /ˌkeɪpə'bɪlətɪ/   *n. (pl. -ies)* ① [C, U] the power or ability to do sth 能力；才能；才干 ② [C, U] the power or weapons that a country has for war or for military action 军事力量；军事武器

recommendation /ˌrekəmen'deɪʃ(ə)n/   *n.* ① [C] an official suggestion about the best thing to do 正式建议；提议 ② [U, C] the act of telling sb that sth is good or useful or that sb would be suitable for a particular job, etc. 推荐；介绍 ③ [C] a formal letter or statement that sb would be suitable for a particular job, etc. 推荐信；介绍信

personalize /'pɜːsənəlaɪz/   *v.* ① (一般作 be personalized) design or produce sth to meet someone's individual requirements 使个性化 ② personify (sth, especially a deity or spirit) 把……人格化（或拟人化）

preference /'prefrəns/   *n.* ① [U, sing.] ~ **(for sb/sth)** a greater interest in or desire for sb/sth than sb/sth else 偏爱；爱好；喜爱 ② [C] a thing that is liked better or best 偏爱的事物；最喜爱的东西

conversion /kən'vɜːʒn/   *n.* [U, C] ~ **(from sth) (into/to sth)** the process or action of changing or causing sth to change from one form to another（使发生）转变；转换；转化

random /'rændəm/   *a.* made, done, happening, or chosen without method or conscious decision 随机的；任意的

blurry /'blɜːri/   *a.* without a clear outline; not clear 模糊不清的

unflattering /ʌn'flætərɪŋ/   *a.* making sb/sth seem worse or less attractive than they really are 贬损的；有损形象的；不恭维的

capture /ˈkæptʃə/	*v.* ① take into one's possession or control by force（用武力）占有；控制；俘获 ② record or express accurately in words or pictures（用文字、图片）正确记录（或表达）
agency /ˈeɪdʒənsɪ/	*n.* ① a business or organization established to provide a particular service, typically one that involves organizing transactions between two other parties 代理；中介；代理处，代办处 ② a department or body providing a specific service for a government or other organization（政府的）办事处；机构
undoubtedly /ʌnˈdaʊtɪdli/	*ad.* without doubt; certainly 毫无疑问地
hallmark /ˈhɔːlmɑːk/	*n.* ① a mark stamped on articles of gold, silver, or platinum certifying their standard of purity（经英国检验机构检验，在金、银或铂金制品上打的）纯度印记 ② a distinctive feature（尤指表示品质优良的）特点；特征
disposal /dɪˈspəʊz(ə)l/	*n.* ① [U] the action or process of throwing away or getting rid of sth 丢弃；摆脱 ② [C] the sale of part of a business, property, etc.（企业、财产等的）变卖；让与
disseminate /dɪˈsemɪneɪt/	*v.* spread or disperse widely 散布；传播
giveaway /ˈɡɪvəweɪ/	*n.* ① a thing that is given free, especially for promotional purposes（尤指促销时商家所给的）赠品 ② a thing that makes an inadvertent revelation 无意中的暴露；露马脚
foster /ˈfɒstə/	*v.* ① encourage or promote the development of (sth, typically sth regarded as good) 鼓励；促进 ② bring up (a child that is not one's own by birth) 收养；领养（非亲生子女）

awareness /əˈweənəs/	*n.* [U, sing.] knowing sth; knowing that sth exists and is important; being interested in sth 知道；认识；意识
participant /pɑːˈtɪsɪpənt/	*n.* a person who takes part in sth 参与者；参加者
definitely /ˈdefɪnɪtli/	*ad.* without question and beyond doubt 一定地；肯定（用于强调）
substitute for	use sth new or different instead of sth else 用……代替
lead to	result in 导致
deter from	discourage someone from doing sth 制止
happen to	chance to be or do sth without intention 碰巧
show off	display proudly; act ostentatiously or pretentiously 炫耀
speaking of	used to introduce a statement or question about a topic recently alluded to 谈起；提到
revolve around	treat as the most important point or element 以……为中心
in the midst of	in the middle of 在……之中；正当……的时候
take over	take on titles, offices, duties or responsibilities 接管；接手
at one's disposal	available for one to use whenever or however one wishes 由某人随意使用；由某人自行支配
apart from	in addition to; as well as 除了……外；也
sign up for	register for; log in 注册；报名
get in touch with	make contact with sb/sth 与……取得联系

Reading Comprehension

Understanding the text

Answer the following questions.

1. What kind of strategy does a fashion brand need to help it stay ahead of its competition?

   A. Global marketing.

   B. Content marketing.

   C. Visual marketing.

   D. Digital marketing.

2. Why is revenue expected to increase when so many fashion stores closed?

   A. Because more than 1,875 fashion stores closed last year.

   B. Because revenue is increasing from £360.7 billion to £534.5 billion.

   C. Because consumers are preferring to shop online rather than visiting a physical location.

   D. Because consumers are getting a website.

3. What if a fashion brand doesn't have a responsive website?

   A. It will lead to superb experience for mobile users.

   B. It will prevent customers from further interactions with the brand.

   C. It will not lead consumers to a competitor.

   D. It will lead users to substitute keyboards for screens.

4. Which of the following is not true in terms of social media marketing for fashion brands?

   A. They can use it to build a relationship with their audience.

   B. They can use certain platforms to promote their products.

   C. They can integrate their store with certain channels like Tencent and Google.

   D. It is an absolute must.

5. Which of the following stategy is not mentioned according to the passage?

   A. Attach an influencer and utilize the latest technologies.

   B. Personalize everything and remarket to your users.

   C. Start a blog and focus on your visuals.

   D. Conduct a market analysis and plan the product.

6. What can be inferred in the last paragraph?

    A. All the marketing strategies mentioned in the passage will be beneficial in the fashion world.

    B. Only one of the marketing strategies in the passage will be beneficial.

    C. You'd better not get in touch with a fashion marketing agency to talk about your project.

    D. None of the above.

Critical thinking

Work in pairs and discuss the following questions.

1. What marketing strategies do you know for fashion brands?

2. What roles do marketing strategies play in fashion brands?

3. How can people achieve success in fashion marketing businesses?

◤ Language Enhancement

Words in use

Fill in the blanks with the words given below. Change the form when necessary. Each word can be used only once.

adapt	preference	approach	contradictory	browse
reside	capture	foster	disseminate	personalize

1. I found the article while I was _____ through some old magazines.

2. All subjects in the curriculum are _____ to the needs of the individual.

3. The world will be different, and we will have to be prepared to _____ to the change.

4. He returned to Denmark in 1930, having _____ abroad for many years.

5. It's fascinating to see how different people _____ the problem.

6. It also provides technical assistance to help countries _____ information on investment opportunities.

7. We give _____ to applicants with some experience.

8. The couple have _____ over 60 children during the past ten years.

9. Of the two _____ aspects, one must be principal and the other secondary.

10. What made him remarkable as a photographer was his skill in _____ the moment.

Expressions in use

Fill in the blanks with the expressions given below. Change the form when necessary. Each expression can be used only once.

substitute for	deter from	attach to	revolve around
take over	at one's disposal	apart from	sign up for

1. You must consider how to better utilize the resources _____.

2. This policy sets the stage for the Department of Commerce to _____ the management of traffic in space.

3. You could _____ ketchup _____ the vinegar according to the book of recipes.

4. _____ a small group of paid staff, the organization consists of unpaid volunteers.

5. Rain showers did not _____ tourists _____ taking in the sights of the National Mall.

6. Many of the arguments about architecture and design _____ misunderstanding what is important.

7. We highly recommend that you _____ courses that will bring you more skills.

8. Professor Hendry said when nicotine is _____ smoke particles it will kill.

Sentence structure

I. Complete the following sentences by translating the Chinese into English, using "prefer to ... rather than ..." structure.

Model: People _____

_____（宁愿去

旅游也不愿假期时待在家里）for a couple of reasons.

→ People prefer to go traveling rather than stay at home during holidays for a couple of reasons.

1. Many college students _____

_____（宁愿听从指令也不愿自己作决定）.

2. The director in the company _____

_____（宁愿工作也不愿坐在那里无所事事）.

3. Why do more and more people _____

_____（宁愿当丁克也不愿生孩子）?

II. Rewrite the following sentences by using "What's important to note here is that ..." structure.

Model:	It is worth noting here that the flu epidemic has not been completely controlled.
	→What's important to note here is that the flu epidemic has not been completely controlled.

1. It is worth noting here that the activity is of great significance to expanding overseas business of our products.

_____
_____.

2. It is worth noting here that smoking is harmful to our health.

_____
_____.

3. It is worth noting here that global warming is posing a growing threat.

_____
_____.

Six Fashion Brands Using Video Marketing Effectively

1　In the world of fashion, you're either in or you're out. At least that's what Heidi Klum tells us on *Project Runway*. What, you don't watch it too? Heidi's not talking about video marketing but if we had a say, we sure would be!

2　By its nature fashion is a visual brand. Customers want to see and feel the brands they love and the styles they crave. Video marketing is a fantastic medium for brands looking to make that type of connection online.

3　If you're in the fashion world, your preferred move is using video marketing to put awesome content in front of your target audience. Some of the biggest brands in fashion are already on this train so there's no time to waste!

4　But don't just take our word for it. Here are a few brands that are crushing their video marketing strategy.

Christian Dior

5　Christian Dior's video marketing content is heavily focused on setting a carefully crafted and artistic scene that conveys their brand's identity.

6　In fact, most of their content transcends the traditional video advertisement and becomes almost like a mini-story themselves. They're visually appealing and beautifully made. Like many high-fashion brands, Dior's video content does a great job of elevating the brand and showing off the style and culture they represent.

Old Navy

7　Wait, what? We go from Christian Dior to Old Navy? True, it's not exactly the highest form of fashion, but Old Navy is a great, fun brand that really knows how to use their video marketing effectively. They not only create great content but also nail it on consistency.

8   A quick look at Old Navy's YouTube channel is all you need to know about the importance of quality content and consistent posting. A closer look shows that Old Navy averages around 20 videos each month.

9   The content itself runs the gamut from promoting sales to advertising new features and all points in between. The videos themselves are not overly complicated or showy for the most part. What they display is effective. Like this quick video that promotes in-store pickup of online orders.

Levi's

10   Levi's is one of the most popular jean companies in the world and has been over 100 years. Their video content is a hodgepodge of both advertisements and content marketing gold. A look at their channel shows they aren't skimping on consistency either. Levi's posts a video every few days and several each week. At its best, video marketing is best approached from multiple angles. Sure, you need to make effective video ads. However, don't neglect the style guides, how-to's and other helpful video types. They go a long way towards raising awareness of your brand and showing your brand's personality.

Old Spice

11   Old Spice has been killing it with their video marketing for a long time. Their content tends to tell a brief story, often in a ridiculous or fantastic setting that is a little over the top but good-natured and fun for all.

11   As an 80-year-old brand, Old Spice has demonstrated an incredible ability to reinvent itself as needed and to engage their target customers in new and innovative ways. This includes response videos where the actor from the above commercial, Isaiah Mustafa, responds to fan comments.

JC Penney

12   JC Penney's video marketing approach is similar to Levi's in that they're killing it from a content marketing perspective. JC Penney's video content is chock full of seasonal ideas, fashion tips, and more.

13   They're also consistent with their posting, adding a new video to their YouTube channel at least once a week and sometimes more. JC Penney, like Levi's is doing a good job of showing their brand identity while also providing helpful information to their viewers.

14   Style guides and fashion tips always make for solid video marketing content as they're much more likely to be shared by your audience than a traditional advertisement.

Chanel

15   Let's bookend it with another luxury brand: Chanel. Chanel also captures the essence of how powerfully effective a helpful approach to video marketing can be. Since cosmetics are a visual medium, it's no surprise that a lot of Chanel's content is tutorial-driven.

16   Chanel also features a variety of celebrities modeling and showcasing their products. Like many luxury brands, Chanel also produces full-length advertisements that don't skimp on style or substance. However, this ad below with Kristen Stewart goes a long way towards showing off the "power of the product" by placing it on a familiar face and showing off the product in action.

Are You Fashion Forward?

17   Fashion brands of all shapes and sizes are making effective use of video marketing online. And for good reason too. Over 40% of YouTube users turn to video content to learn more about a product.

18   When it comes to fashion and style, video marketing is simply the most effective tool for reaching and engaging with customers online. The brands

above, as well as many others, are truly taking advantage of this medium by producing consistent, quality video content.

19   If you're a fashion brand looking to make a splash, video marketing is the way to go. Create beautiful, professional-quality video ads that tell your brand story, showcase your style, and perhaps most importantly: demonstrate just how great your products are.

Unit 6

Diversity and Inclusion in Fashion

Fashion is just an epidemic.

—*George Bernard Shaw*

Everyone laughs at the old fashion, but devoutly pursues the novel fashion.

—*Henry David Thoreau*

Pre-Reading Activities

1. Listen to a short talk about diversity and inclusion in fashion and fill in the blanks based on what you hear.

(1) _____ is one of the biggest industries in the world that capitalizes on (2) _____ and self-expression. Fashion isn't all about (3) _____ and garments; it's an industry that prides itself in selling self-esteem and self-identity. Without diversity and inclusion, fashion cannot truly (4) _____ as an art form. The industry has a long-standing (5) _____ of discriminating against race, size, gender, and (6) _____, often depicting the "perfect model" as tall, fair, and slender. Due to this old way of thinking, the industry needs reformation in terms of inclusivity and diversity. But to (7) _____ accountability, there must be clarity in what those two words mean for the industry. The words "inclusion" and "diversity" are often used interchangeably without a clear definition. We must understand and (8) _____ on the definition of each word.

2. Discuss the following questions with your partner.

(1) Do you think the fashion industry is becoming more diverse?

(2) What do you know about fast fashion?

(3) What are the characteristics of fashion industry in modern times?

The All-inclusive New Era in Fashion

1   The fashion industry has been quick to **embrace** trends yet not so quick to embrace inclusivity. As time has progressed, there has been a **gradual** shift in the fashion **scene adopting** diversity within its marketing and branding strategy. Years ago, seeing a **transgender** or plus size model **featured in mainstream** fashion was very unlikely. However, high fashion and high street fashion brands have shown a **progressive** marketing approach through adopting strategies which speak to their multicultural, multiracial and gender **fluid** consumers.

2   The world of fashion has entered a new **era** of breaking **boundaries, in answer to** growing challenges from social media **exposure**, consumer **activism**, and increased public socio-political awareness. Brands have faced new pressures to **showcase** their willingness to be more socially inclusive than previous fashion generations.

3   **Retailers** and brands have become increasingly open to targeting a wider globalized market through **catering to** a more **diverse** and **opinionated** customer base. This has, however, **coincided with** the changing cultural and political field in the world of social media.

4   Racially and culturally **insensitive** campaigns and designs exhibited by luxury fashion brands has proven to be **detrimental** to the brands' public image, in turn **negatively** effecting sales. In particular, the storm of **outspoken** and **dominant** influencers addressing such issues to wider audiences has placed great stress on the fashion industry to embrace diversity, which has traditionally been avoided in the past.

5   In this sense, social media has totally shifted the fashion scene by giving different **minorities**—usually ignored by the fashion world—a visual

platform and a public voice. Public apologies for **inappropriate** and **controversial** moves have become a common response by many fashion brands, through social media accounts, being held accountable for every little move on social media.

6   Social media has encouraged a culture of public shaming, this has become an easy means for social activists to publicly **criticize** and **scrutinize** marketing methods of fashion brands, big or small. The Internet has also proven to be an effective platform for strong and active voices which have played a **significant** role in directing the industry towards diverse marketing and branding.

7   Retailers have used **inclusive** marketing to **tap into** a number of previously neglected markets, one of these being **modest** fashion. The rise of globalization led to a growing increase in modest fashion bloggers **emerging** on social media with the aim of **subverting conventional notions** of woman exhibited by the mainstream media.

8   Some Brands have showcased modest fashion collections, and other consumers who choose to wear higher necklines and increased **hemlines**— seemingly more and more common in mainstream fashion.

9   Luxury e-commerce businesses such as Net-a-Porter and Farfetch have also been quick to move into the modest fashion market. In 2016, Yoox Net-a-Porter Group completed a joint venture with the aim to create an unprecedented luxury e-commerce platform within the Middle East.

10   Two years later, Farfetch struck up a partnership with modest luxury retailer The Modist, allowing Farfetch to showcase a series of modest fashion online and **exclusive** pieces for the website and from The Modist's own brand Layeur, as well as one of designs from other labels. Such **investments** from the luxury fashion world mark a new era of social inclusivity which in the past was considered out of the ordinary.

11   For a long time, plus sizing had not been **accessible** for many female consumers, but plus-size and **curvy** clothing has rapidly become

incorporated into mainstream fashion in the last few years. Fast fashion retailers are popular brands providing many pieces in **extended** sizes. On the opposite end of the fashion **spectrum**, high-end designers have become more and more willing to include plus-size models on the **runway**.

12   As a result, plus-size influencers such as American model and body activist Ashley Graham has used her presence on the **catwalk** to encourage inclusivity within fashion and fashion branding. Such public figures have led more and more brands to include extended sizes as part of their collections. Most recently, the Spanish bridal wear brand Pronovias announced an exclusive collaboration with Ashley Graham to cater to curvier and plus-size brides. Seeing a plus-size female as a main figure of a fashion marketing campaign, on the front cover of a fashion magazine, or even walking down the high-end runway was previously unheard of; the power of **influential** plus-size bloggers, models and body activists has forced the fashion industry to become more inclusive.

13   In addition to an inclusive shift within fashion marketing and branding, we have also approached a generation of a more visible shift across the fashion industry as a whole. 2017 was a **pivotal** year: British *Vogue* employed the very first black editor, Edward Enninful (who has not wasted time making up for the past lack of diversity, in particular on its covers). Virgil Abloh has also become the first black creative director at LVMH.

14   Key leaders in the fashion industry are focusing their attention on being more diverse within the workspace—with luxury designers such as Burberry, Gucci and Chanel recently employing diversity and inclusion officers—**reflecting** the changing social **landscape** of fashion branding and marketing.

Notes

Farfetch   英国著名时尚购物电商平台

Yoox Net-a-Porter Group   全球最大在线时尚奢侈品零售商

Middle East　中东

The Modist　中东奢侈品零售电商

Pretty Little Thing　英国 Boohoo 旗下的新兴女装时尚品牌

Ashley Graham　阿什利·格雷厄姆，美国第一大码超模

Pronovias　普洛诺维斯，西班牙国宝级婚纱礼服品牌

Vogue　知名综合性时尚生活类杂志

New words and phrases

embrace /ɪmˈbreɪs/	*v.* ① hold (sb) closely in one's arms, especially as a sign of affection 拥抱；怀抱 ② accept or support (a belief, theory, or change) willingly and enthusiastically 欣然接受；信奉
	n. ① [C] an act of holding someone closely in one's arms 拥抱，怀抱 ② [U] act of accepting or supporting sth willingly or enthusiastically 欣然接受（或支持）
gradual /ˈɡrædʒuəl/	*a.* taking place or progressing slowly or by degrees （发生或发展）缓慢地；逐渐地
scene /siːn/	*n.* ① the place where an incident in real life or fiction occurs or occurred 发生地点；现场 ② an event or a situation that you see, especially one of a particular type 事件；场面；情景 ③ [sing.] a specified area of activity or interest 活动领域；界；坛；圈子 ④ a sequence of continuous action in a play, film, opera, or book （戏剧、电影等中的）一段情节
adopt /əˈdɒpt/	*v.* ① legally take (another's child) and bring it up as one's own 收养 ② take up or start to use or follow (especially an idea, method, or course of action) 采用；采纳

transgender /trænz'dʒendə/   *a.* relating to transsexuals and transvestites 变性（者）的；易性癖（者）的；异装癖（者）的

feature /'fi:tʃə/   *v.* have as a prominent attribute or aspect 以⋯⋯为特点（或特征）

n. ① a distinctive attribute or aspect of sth 特色；特征 ② a newspaper or magazine article or a broadcast program devoted to the treatment of a particular topic, typically at length 特写；专题报道

mainstream /'meɪnstri:m/   *n.* [sing.] the ideas, attitudes, or activities that are regarded as normal or conventional; the dominant trend in opinion, fashion, or the arts 主流；主要倾向

progressive /prə'gresɪv/   *a.* happening or developing gradually or in stages; proceeding step by step 进步的；先进的；逐步的；渐次的

fluid /'flu:ɪd/   *n.* ① [C, U] a liquid; a substance that can flow 流体，液体；液

a. ① able to flow easily 能流动的；易流动的 ② not settled or stable; likely or able to change 不稳定的；易变的 ③ smoothly elegant or graceful 优美的；流畅的

era /'ɪərə/   *n.* a long and distinct period of history with a particular feature or characteristic 时代

boundary /'baʊndri/   *n.* ① a line which marks the limits of an area; a dividing line 分界线，边界；界限 ② (usually *pl.*) a limit of sth abstract, especially a subject or sphere of activity（某些抽象事物的）界限

exposure /ɪk'spəʊʒə/   *n.* ① [U] the state of having no protection from contact with sth harmful 暴露（不受保护的接触）② [U] the publicizing of information or an event 披露；公布

activism /'æktɪvɪzəm/　　　　*n.* [U] the policy or action of using vigorous campaigning to bring about political or social change 行动主义（或行为）；激进主义（或行为）

showcase /'ʃəʊkeɪs/　　　　*v.* exhibit; display 展示；表现

n. ① a glass case used for displaying articles in a shop or museum（商店或博物馆的）玻璃陈列柜 ② a place or occasion for presenting something favorably to general attention 展示场所（或场合）

retailer /'riteɪlə/　　　　*n.* [C] a merchant who sells goods at retail 零售商

diverse /daɪ'vɜːs/　　　　*a.* showing a great deal of variety; very different 多样的；相异的

opinionated /ə'pɪnjəneɪtɪd/　　　　*a.* having very strong opinions that you are not willing to change 固执己见的；顽固的

coincide /ˌkəʊɪn'saɪd/　　　　*v.* ① occur at or during the same time 同时发生 ② be in agreement 一致；协调

insensitive /ɪn'sensətɪv/　　　　*a.* ① not sensitive to a physical sensation 不敏感的 ② not aware of or able to respond to sth 无反应的；感觉迟钝的 ③ showing or feeling no concern for others' feelings 漠不关心的；无情的

detrimental /ˌdetrɪ'mentl/　　　　*a.* tending to cause harm 有害的；不利的

negatively /'negətɪvli/　　　　*ad.* in a harmful manner or a negative way 消极地；负面地；否定地

outspoken /aʊt'spəʊkən/　　　　*a.* frank in stating one's opinions, especially if they are shocking or controversial 坦率的；直言的

dominant /'dɒmɪnənt/　　　　*a.* most important, powerful, or influential（最）重要的；（最）强大的；（最）有影响的

inappropriate /ɪnə'prəʊprɪət/　　　　*a.* not suitable or proper in the circumstances 不恰当的；不相宜的

controversial /ˌkɑːntrə'vɜːʃl/　　　　*a.* giving rise or likely to give rise to public disagreement 引起争论的；有争议的

criticize /'krɪtɪsaɪz/   v. ① indicate the faults of (sb or sth) in a disapproving way 批评；责备 ② form and express a sophisticated judgment of (a literary or artistic work) 评论；评价（文艺作品）

scrutinize /'skru:tənaɪz/   v. examine or inspect closely and thoroughly 细查

significant /sɪg'nɪfɪkənt/   a. ① sufficiently great or important to be worthy of attention; noteworthy 重要的；显著的 ② suggesting a meaning or message that is not explicitly stated 意味深长的

inclusive /ɪn'klu:sɪv/   a. ① including a wide range of people, things, ideas, etc. 包容广阔的；范围广泛的 ② ~ (of sth) having the total cost, or the cost of sth. that is mentioned, contained in the price 包含全部费用

modest /'mɒdɪst/   a. ① unassuming or moderate in the estimation of one's abilities or achievements 谦虚的；谦逊的 ② relatively moderate, limited, or small 适中的；适度的；有限的 ③ shy about showing much of the body; not intended to attract attention 庄重的；朴素的

emerge /ɪ'mɜ:dʒ/   v. ① move out of or away from sth and come into view 浮现；显露 ② become apparent, important, or prominent 变得显眼（或重要、突出）

subvert /səb'vɜ:t/   v. undermine the power and authority of (an established system or institution) 颠覆；破坏

conventional /kən'venʃənl/   a. ① based on or in accordance with what is generally done or believed 传统的；习惯的 ② (of a person) concerned with what is generally held to be acceptable at the expense of individuality and sincerity 因循守旧的；墨守成规的

hemline /'hemlaɪn/   n. [C] the level of the lower edge of a garment such as a skirt, dress, or coat （衣裙的）底边；下摆

exclusive /ɪk'sklu:sɪv/   *a.* ① excluding or not admitting other things 排他的 ② restricted or limited to the person, group, or area concerned 独有的；独享的；专用的

n. [C] an item or story published or broadcast by only one source 独家新闻；独家报道

investment /ɪn'vestmənt/   *n.* ① [U] the action or process of investing money for profit or material result 投资 ② [C] a thing that is worth buying because it may be profitable or useful in the future 值得投资之物 ③ [C, U] an act of devoting time, effort, or energy to a particular task in order to make it successful（时间、精力的）投入；付出

accessible /ək'sesəbl/   *a.* ① able to be reached or entered 可到达的；可进入的 ② able to be easily obtained or used 易得到的；易使用的

curvy /'kɜ:vɪ/   *a.* ① having curves 有曲线的；弯曲的 ② (especially of a woman's figure) shapely and voluptuous（尤指女子身材）曲线美的

extended /ɪk'stendɪd/   *a.* long or longer than usual or expected 延长了的；扩展了的

spectrum /'spektrəm/   *n.* (*pl.* spectra) ① a band of colors, as seen in a rainbow, produced by separation of the components of light by their different degrees of refraction according to wavelength 光谱 ② used to classify sth, or suggest that it can be classified, in terms of its position on a scale between two extreme or opposite points 范围

runway /'rʌnweɪ/   *n.* ① a strip of hard ground along which aircraft take off and land（机场的）跑道 ② a raised gangway extending into the audience in a theatre or other public venue, especially as used for fashion shows 延伸台道（尤指用作时装表演的延伸台道）

catwalk /'kætwɔːk/   *n.* ① a narrow walkway or open bridge, especially in an industrial installation 狭窄通道或天桥 ② a platform extending into an auditorium, along which models walk to display clothes in fashion shows 时装表演台；T 型舞台

influential /ˌɪnflʊ'enʃl/   *a.* having great influence on sb or sth 有影响的；有势力的

pivotal /'pɪvətl/   *a.* of crucial importance in relation to the development or success of sth else （对事物的发展或成功）起关键性作用的；起中心作用的

reflect /rɪ'flekt/   *v.* ① (of a surface or body) throw back (heat, light, or sound) without absorbing it 反射（热、光或声音）② embody or represent sth in a faithful or appropriate way 忠实反映；恰当表明 ③ ~ **on/upon** think deeply or carefully about 深思；反省

landscape /'lændskeɪp/   *n.* ① [C, usually sing.] all the visible features of an area of countryside or land, often considered in terms of their aesthetic appeal 风景；风光 ② [C] the distinctive features of a particular situation or intellectual activity 局势；局面

feature in   be a significant characteristic of or take an important part in 是……的重要特点；在……中起重要作用

in answer to   as a response to or as a result of 作为对……响应

cater to   try to satisfy (a particular need or demand) 迎合；满足需要

coincide with   correspond in nature; tally 一致；相符；吻合

tap into   make use of a source of energy, knowledge, etc. that already exists 利用；开发；发掘

Reading Comprehension

Understanding the text

Answer the following questions.

1. Why does the author argue that high fashion and high street fashion brands have shown a progressive way of marketing?

2. What pressures have fashion brands faced in a new era?

3. How have retailers and brands become increasingly open?

4. Why has the Internet been an effective platform for strong and active voices?

5. What did the rise of globalization and terrorism bring to fashion marketing?

6. Why did Farfetch begin a partnership with luxury retailer The Modist?

7. What has rapidly become mainstream fashion in the last few years?

8. What enabled Ashley Graham to encourage inclusivity in fashion branding?

9. What have happened apart from an inclusive change across the fashion industry?

10. What are officers in the fashion industry focusing their attention on?

Critical thinking

I. Make a Presentation.

Choose two or three fashion stores in a shopping mall to carry out a survey to learn about the characteristics of the latest fashion style in each store. The survey may include the information about the diversity and inclusion in fashion. You are supposed to be honest about the information. Work in groups and summarize what you have discussed in a report, then present the report to the class.

II. Work in pairs and discuss the following questions.

1. What changes have taken place in the new era of fashion world?

2. How does fashion relate to globalization?

3. What pressures do you think fashion brands are facing?

◼ Language Enhancement

Words in use

Fill in the blanks with the words given below. Change the form when necessary. Each word can only be used once.

embrace	adopt	launch	trigger	diverse
emerge	dominant	controversial	criticize	accessible

1. Immigration is regarded as a _____ issue in some countries all the time.
2. An allergy can be _____ by stress or overwork.
3. The opposition _____ the government's failure to consult adequately.
4. The company is now in an even more _____ position in the market.
5. Online courses are so convenient and popular that learning opportunities are made more _____ to adults.
6. The government is to _____ a £1.25 million publicity campaign.
7. After a long separation, Peter held her beloved girl in a warm _____ when they meet again.
8. This approach has been _____ by a few large-scale science and research institutes.
9. _____ subjects such as architecture, language teaching, and the physical sciences are arranged in the curriculum, which is very flexible.
10. Reports of a deadlock _____ during preliminary discussions.

Banked cloze

Fill in the blanks by selecting suitable words from the word bank. You may not use any of the words more than once.

A. respond	F. challenges	K. profitable
B. potential	G. emerged	L. representing
C. diverse	H. approaches	M. attempts
D. delivered	I. adequate	N. divided
E. resent	J. presenting	O. garments

Closely related to marketing is merchandising, which 1. _____ to maximize sales and profitability by inducing consumers to buy a company's products. In the standard definition of the term, merchandising involves selling the right product, at the right price, at the right time and place, to the right customers. Fashion merchandisers must thus utilize marketers' information about customer preferences as the basis for decisions about such things as stocking appropriate merchandise in 2. _____ but not excessive quantities, offering items for sale at attractive but still 3. _____ prices, and discounting overstocked goods. Merchandising also involves 4. _____ goods attractively and accessibly through the use of store windows, in-store displays, and special promotional events. Merchandising specialists must be able to 5. _____ to surges in demand by rapidly acquiring new stocks of the favored product. An inventory-tracking computer program in a department store in London, for example, can trigger an automatic order to a production facility in Shanghai for a certain quantity of 6. _____ of a specified type and size to be 7. _____ in a matter of days.

By the early 21st century the Internet had become an increasingly important retail outlet, creating new 8. _____ (e.g., the inability for customers to try on clothes prior to purchase, the need for facilities designed to handle clothing returns and exchanges) and opening up new opportunities for merchandisers (e.g., the ability to provide customers with shopping opportunities 24 hours per day, affording access to rural customers). In an era of increasingly 9. _____ shopping options for retail customers and of intense price competition among retailers, merchandising has 10. _____ as one of the cornerstones of the modern fashion industry.

Expressions in use

Fill in the blanks with the expressions given below. Change the form when necessary. Each expression can be used only once.

feature in	in particular	capitalize on	strike up
tap into	in answer to	coincide with	make up for

1. If you _____ a friendship with someone, that means you begin to have a real and sincere emotion with him or her.
2. I hurried along the passage _____ the doorbell's ring.
3. No matter what he did, he couldn't _____ the gap in their friendship.
4. Is there anything _____ you'd like for dinner?
5. Experimental results _____ theoretical calculations quite well.
6. The latest movies from British directors _____ the current film season.
7. Some cadres who _____ their functions and powers to encroach on state or collective property must be severely punished.
8. The success people tend to be those who are best at _____ their own potential.

◼ Translation

I. Translate the following paragraph into Chinese.

The fashion industry forms part of a larger social and cultural phenomenon known as the "fashion system," a concept that embraces not only the business of fashion but also the art and craft of fashion, and not only production but also consumption. The fashion designer is an important factor, but also is the individual consumer who chooses, buys, and wears clothes, as well as the language and imagery that contribute to how consumers think about fashion. The fashion system involves all the factors that are involved in the entire process of fashion change. Some factors are intrinsic to fashion, which involves variation for the sake of novelty (e.g., when hemlines have been low for a while, they will rise). Other factors are external (e.g., major historical events such as wars, revolutions, economic booms or busts, and the feminist movement). Individual trendsetters (e.g., Madonna and Diana, princess of Wales) also play a role, as do changes in lifestyle (e.g., new sports, as when skateboarding was introduced in the 1960s) and music (e.g., rock and roll, hip-hop).

II. Translate the following paragraph into English.

时装设计师和制造商不仅向零售商推销他们的服装，而且向媒体和顾客直接推销他们的服装。早在 19 世纪末，巴黎时装店就开始为顾客提供最新时装的私人观赏服务。到了 20 世纪初，不仅时装店，百货公司也定期举办有专业模特参加的时装秀。模仿巴黎的时装设计师，其他国家的成衣设计师也开始为私人客户、记者和买家举办时装秀。在 20 世纪末和 21 世纪初，时装表演变得更加精致和戏剧化，在更大的场所举行，为模特们特别搭建 "T 型台"，并在展示新时尚方面发挥越来越突出的作用。

Paragraph Writing

How to develop a comparison/contrast essay

You often compare and contrast two things, items, or people in everyday life. To compare means to show the similarities and to contrast means to show the differences. In everyday life, you compare and contrast when you buy a car or choose an apartment. You can also compare two teachers, two jobs, or two ways to spend a weekend. There are some important points to remember when you write a comparison/contrast essay:

1. The items being compared or contrasted must be from the same category. In other words, they should be comparable. For example, you can compare a desktop with a laptop, but you may not compare a computer with a camera.

2. It is better to mention the points you would like to compare or contrast in the thesis statement so as to let readers know about the main idea in your essay right away. For example, to compare two jobs, you may want to write a thesis statement like this: *It's not difficult to choose an ideal job if you compare the job features of salary, benefits, and opportunity for promotion.*

3. There are two basic patterns to organize the details of your essay: **point-by-point** and **subject-by-subject**. Here, *point* refers to the aspects you would like to use to develop your essay, and *subject* refers to the two items or people you want to compare or contrast. In this unit, you will learn how to use the point-by-point pattern. For instance, if you compare two generations, you may organize the details as follows:

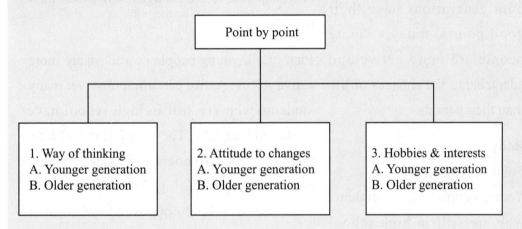

4. The key to write a successful and coherent essay is the appropriate use of comparison and contrast structure words. The common ones are:

Comparison: *similarly, likewise, also, too, as, just as, like, just like, similar to, alike, the same as, not only ... but also..., both ... and ..., etc.*

Contrast: *on the other hand, in contrast, however, in comparison, even though, although, though, whereas, while, but, different from, unlike, etc.*

Structured writing

Read the sample essay and see how the comparison/contrast essay is developed.

Topic

The differences between the young generation and their parents' generation

Introduction

Thesis statement: Even though both generations have their good points, today's young people are more active and adaptable to the changes of life than their parents.

Body

Point 1: active in life

Young people: Many students who are still in high school have started to find jobs.

Parents: They often waited until they were qualified enough to get a job.

Point 2: adaptable to the changes of life

Young people: find the Internet useful to their studies and work.

Parents: find it hard to remember how to use computers or surf online

Sample essay

As is known to all, people's lifestyles and behavior change constantly. Consequently, there are many differences between the young generation and their parents' generation. Even though both generations have their good points, today's young people are more active and adaptable to the changes of life than their parents.

First of all, young people nowadays are more active in life. As we can often observe, many students who are still in high school have started to find jobs. They think this will help them gain more experience and they can work better after graduating from universities. This is an absolutely opposite view to their parents' generation. In their times, parents often waited until they were qualified enough to get a job. They were not as active as young people today.

In addition, the young generation better adapts to the changes of life. For example, for the past decade or so, the Internet has developed tremendously. Young people find it extremely easy to use and feel very comfortable to apply it to their studies and work; therefore, they have made more progress at school and at

Conclusion

Young people usually live in a more active and flexible way than their parents.

work. In contrast, their parents usually find it hard to remember how to use computers or surf online. They often get frustrated when they have to work on a computer and find it hard to adapt to the new technology.

To conclude, although every generation has its own uniqueness and abilities to live their lives, young people usually live in a more active and flexible way than their parents.

Write an essay of no less than 150 words on the topic "Generation Gap Between Parents and Children". You can follow the outline given below.

Topic:

Generation Gap Between Parents and Children

Introduction:

Thesis statement: The differences between parents and children can be well observed in their completely different attitudes and views.

Body:

Point 1: Differences in their attitude toward life
Point 2: Differences in the way to look at things

Conclusion:

It's very difficult for parents and their youngsters to get along due to their distinct attitudes toward life and the way to view things.

More topics:

High School vs. University

My Two Best Friends

Why Is Everyone Talking About Fast Fashion?

1   According to *The Wall Street Journal*, the average person will buy 68 garments this year, each of which is worn only seven times before **disposing of** it. It wasn't always like this. In 1980, people bought five times fewer pieces of clothing, and kept them much longer—but the rise of fast fashion has drastically changed the clothing industry, **flooding** the market **with** cheap, poorly-made garments.

2   What used to be a rare purchase, **thoughtfully** considered, has turned into a series of **impulse** buys made **at** ever-shorter **intervals**. Consumers can pick up a new fast fashion brand dress each week without a second thought, then **toss** it after **donning** it once or twice. And while it's great to see good style **democratized**, this new business model has had **disastrous** effects.

3   Fast fashion **refers to** clothing manufactured at high speed and sold at a low price point. Companies in the fast fashion game sell very cheap clothes. And rather than **releasing** two or more collections at certain times each year, as was the standard in the fashion industry for generations, they constantly push out new product, enabling them to **keep pace with** rapidly evolving consumer tastes. Famously, Fashion Nova takes this to the Nth degree, **rolling out imitations** of Kardashian-approved looks just a day or two after Kim Kardashian **debuts** a new dress on Instagram.

4   This business model wasn't born on social media, but the Internet has certainly **exacerbated** the problem. Dana Thomas, a veteran style reporter and the author of *Fashionopolis: The Price of Fast Fashion and the Future of Clothes*, **traces** the rise of fast fashion to the late 1980s. In the decades since, we've seen clothes decrease in quality as they increase in quantity.

5   Social media has only **accelerated** the problem. The desire to never be photographed in the same outfit twice, combined with the constant advertising—both in influencer's **sponsored** posts and actual ads—**interspersed** in users' Instagram feeds, fuels a desire for constant wardrobe **renewal**. As one teen recently told *The New York Times*, "I wouldn't really want someone seeing me in a dress more than once. People might think I didn't have style if I wore the same thing over and over."

6   The industry has faced frequent criticism for its outsize environmental impact. There are some very real **ecological** costs **associated with** these bargain price tags—and in recent years, fast fashion's environmental **toll** has only increased. **According to** the United Nations Environment Program, 10% of global greenhouse gas **emissions attribute to** the clothing industry—more **aviation** and shipping combined, as per the *Los Angeles Times*. It's also **responsible for** a large portion of water pollution worldwide, and uses **copious** chemicals.

7   Of the clothes produced, **approximately** 20% are never purchased, and quickly find their way to **landfills**. More garments than ever are made with **synthetic** fabrics that don't naturally decompose, **compounding** the waste problem.

8   Brands in the fast fashion space are also often caught **exploiting** workers, both internationally and **domestically**. Fast fashion's low price points rely on even lower manufacturing costs. To keep margins as high as possible, brands outsource production to companies all over the world in search of cheap labor. This poorly regulated supply chain has created unacceptable working conditions for people around the globe. For many, the 2013 **collapse** of the Rana **Plaza** garment factory in Bangladesh, which killed 1,100 and injured many more, has become a symbol of how bad things have gotten.

9   But this isn't just an international problem. **Contractors** in the US producing clothes for fast fashion companies have been caught paying

employees far below the **minimum** wage. A recent **investigation**, for example, **revealed** that workers creating Fashion Nova clothing in Los Angeles were being paid as little as $2.77 an hour.

10   When a brand is called out for substandard working conditions, they often **claim ignorance**, noting that they **commission** third party companies to produce their products. As a designer told the *Times*, "We don't own the sewing contractor, so whatever the sewing contractor does, that's his problem." Thankfully, there are some alternatives—but they often have their own downsides.

11   Generally speaking, consumers have two ways to avoid supporting fast fashion: buying from more **ethical** brands, or purchasing used clothing. (And either way, the longer you keep something in your closet, the better.)

12   Everlane is often **hailed** for its **transparent** supply chain, but the recent news that the company's customer service team is **unionizing** reveals that even **purportedly virtuous** brands can have their **shortfalls**.

13   The RealReal, an online marketplace for secondhand designer clothing, informs customers about the environmental costs that will be avoided or **offset** by buying each used item. Still, the company, which claims to **authenticate** every designer item, has been repeatedly accused of selling **fakes**.

14   It's never been harder—or more crucial—to be an informed fashion consumer.

Notes

The Wall Street Journal　《华尔街日报》，美国付费发行量最大的财经报纸

Fashion Nova　美国快时尚品牌、电商平台

Kim Kardashian　金·卡戴珊，美国服装设计师、演员

Dana Thomas　达娜·托马斯，美国时尚领域记者

The New York Times　《纽约时报》，一份美国纽约出版的日报

United Nations Environment Program　联合国环境规划署

Los Angeles Times　《洛杉矶时报》，美国西部最大的对开日报

Bangladesh　孟加拉国（位于恒河三角洲）

The Times　《泰晤士报》，英国最大的主流日报

Everlane　艾芙兰公司旗下的一个服饰品牌

The RealReal　美国二手奢侈品寄售网站

New words and phrases

thoughtfully /ˈθɔːtf(ə)li/　*ad.* ① in a thoughtful manner 沉思地 ② showing consideration and thoughtfulness 体贴地

impulse /ˈɪmpʌls/　*n.* ① a sudden strong and unreflective urge or desire to act 冲动 ② a driving or motivating force; an impetus 推动力；冲力

interval /ˈɪntəvl/　*n.* ① an intervening time or space 间隔；间歇 ② a pause; a break in activity 停顿；中止；暂停

toss /tɒs/　*v.* throw sth somewhere lightly, easily, or casually 轻轻地（或随意地）扔、抛、掷

n. an action or instance of throwing sth 扔、掷、抛

don /dɒn/　*v.* put on (an item of clothing) 穿上（一件衣服）

democratize /dɪˈmɒkrətaɪz/　*v.* ① introduce a democratic system or democratic principles to 向……引入民主制度；使民主化 ② make sth accessible to everyone 普及；使大众化

disastrous /dɪˈzɑːstrəs/　*a.* causing great damage 灾难性的

release /rɪˈliːs/　*v.* ① allow or enable to escape from confinement; set free 释放；解放 ② make sth available to the public 公开；公布；发布

① [U] the act of making sth available to the public 公开；发行；发布 ② [C] a thing that is made available to the public, especially a new CD or film/movie 新发行的东西；尤指新激光唱片，新电影

imitation /ˌɪmɪˈteɪʃn/   *n.* ① [C] a thing intended to simulate or copy sth else 仿制品；赝品 ② [U] the act of copying sb/sth 模仿；效仿

debut /deɪˈbjuː/   *n.* ① a person's first appearance or performance in a particular capacity or role 初试锋芒；崭露头角 ② the first public appearance of a new product or presentation of a theatrical show （新产品的）首次亮相；（戏剧的）首场演出

v. ① perform in public for the first time 首场演出；初次登场 ② launch (a new product) 推出（新产品）

exacerbate /ɪgˈzæsəbeɪt/   *v.* make (a problem, bad situation, or negative feeling) worse 使（问题、严峻的形势、消极情感）恶化

trace /treɪs/   *n.* [U, C] a mark, an object, or other indication of the existence or passing of sth 痕迹；踪迹

v. ① find or describe the origin or development of 追溯；追究；查考 ② describe a process or the development of sth 描绘（事物的过程或发展）；追述；记述

accelerate /əkˈseləreɪt/   *v.* ① (of a vehicle or other physical object) begin to move more quickly （车辆或其他物体）加速 ② increase in amount or extent （数量或程度上）增加；增长

v. cause to go faster 使加速；加快

sponsor /ˈspɒnsə/   *n.* a person or organization that provides funds for a project or activity carried out by another, in particular 资助人；赞助者；资助（或赞助）单位

v. ① pay some or all of the costs involved in staging (a sporting or artistic event) in return for advertising 赞助（体育，文艺活动）② pledge to donate a certain sum of money to (sb) after they have participated in a fund-raising event organized on behalf of a charity 承诺向……捐赠，承诺赞助

intersperse /ˌɪntəˈspɜːs/   *v.* ① scatter among or between other things; place here and there 散布；散置；穿插 ② diversify (a thing or things) with other things at intervals 点缀

renewal /rɪˈnuːəl/   *n.* ① [C] an instance of resuming an activity or state after an interruption 重新开始；恢复；继续 ② [U, C] the replacing or repair of sth that is worn out, run-down, or broken 更换；更新；重建；复兴

ecological /iːkəˈlɒdʒɪk(ə)l/   *a.* ① characterized by the interdependence of living organisms in an environment 生态的；生态学的 ② interested in and concerned about the ecology of a place 关注生态环境的；主张生态保护的

toll /təʊl/   *v.* charge a toll for the use of (a bridge or road) 收取（桥梁或道路的）通行费

n. ① [C] a charge payable for permission to use a particular bridge or road （桥梁或道路的）通行费 ② [U] the cost or damage resulting from something （付出的）代价；（遭受的）损失，破坏

emission /ɪˈmɪʃn/   [U] the production and discharge of sth, especially gas or radiation 发出；散发；排放

aviation /ˌeɪviˈeɪʃn/   *n.* [U] the flying or operating of aircraft 飞行；航空

copious /ˈkəʊpɪəs/   *a.* abundant in supply or quantity 丰富的；充裕的

approximately /əˈprɒksɪmətlɪ/   *ad.* (of quantities) imprecise but fairly close to correct 大约

landfill /'lændfɪl/

n. ① [U, C] an area of land where large amounts of waste material are buried under the earth 废物填埋地（或场） ② [U] the process of burying large amounts of waste material 废物填埋 ③ [U] waste material that will be buried 填埋的废物

synthetic /sɪn'θetɪk/

a. made by chemical synthesis, especially to imitate a natural product 合成的；人造的

n. [usually *pl.*] a synthetic material or chemical, especially a textile fibre 合成物（尤指合成纤维）

compound /'kɒmpaʊnd/

a. made up or consisting of several parts or elements, in particular 复合的；混合的；合成的；化合的

n. a thing that is composed of two or more separate elements; a mixture of two or more things 复合物；混合物

v. ① make up；constitute 使复合；使合成；组成 ② make sth bad worse; intensify the negative aspects of 使恶化；使更糟；使复杂化

exploit /ɪk'splɔɪt/

v. ① make full use of and derive benefit from 充分利用 ② benefit unjustly or unfairly from the work of sb typically by overworking or underpaying them 剥削

domestically /də'mestɪkli/

ad. ① with respect to the internal affairs of a government 国内地 ② with respect to home or family 家庭地；家庭式地

collapse /kə'læps/

v. ① (of a structure) fall down or in; give way（建筑）倒塌；塌陷 ② (of a person) fall down and become unconscious, typically through illness or injury 昏倒 ③ (of an institution or undertaking) fail suddenly and completely 突然失败；崩溃；倒闭

n. ① an instance of a structure falling down or in 倒塌；塌陷 ② a sudden failure of an institution or undertaking 突然倒闭；崩溃；瓦解 ③ a physical or mental breakdown 崩溃；垮掉

plaza /'plɑːzə/   *n.* ① a public square, marketplace, or similar open space in a built-up area 广场 ② a shopping centre 购物中心

contractor /kən'træktə/   *n.* a person or firm that undertakes a contract to provide materials or labor to perform a service or do a job 订约人；承包人；承包商

minimum /'mɪnɪməm/   *a.* smallest or lowest 最小的；最少的；最低的

n. the least or smallest amount or quantity possible, attainable, or required 极小值；最低限度；最低点

investigation /ɪnˌvestɪ'geɪʃn/   *n.* ① [C, U] the action of investigating sth or sb; formal or systematic examination or research 调查；审查 ② [C, U] a formal inquiry or systematic study 调查研究

reveal /rɪ'viːl/   *v.* ① make (previously unknown or secret information) known to others 透露；暴露；泄露；揭露 ② cause or allow sth to be seen 使显露；展现；显示

claim /kleɪm/   *v.* ① state or assert that sth is the case, typically without providing evidence or proof 声称；断言 ② formally request or demand; say that one owns or has earned 要求；索取；认领

n. ① an assertion of the truth of sth, typically one which is disputed or in doubt 主张；宣称；断言 ② a request for sth considered one's due from a company, the government, etc. 索款；索赔

ignorance /'ɪgnərəns/   *n.* [U] lack of knowledge or information 无学；无知

commission /kəˈmɪʃn/   *n.* ① an instruction, command, or role given to a person or to a specially constituted group 委托；委任；授权；任命；任务 ② a group of people entrusted by a government or other official body with authority to do sth 委员会 ③ a sum, typically a set percentage of the value involved, paid to an agent in a commercial transaction 佣金；回扣

v. give an order for or authorize the production of sth, such as a building, piece of equipment, or work of art 授权（或委托）制作（某物，如建筑物、设备或艺术品）

ethical /ˈeθɪkl/   *a.* of or relating to moral principles or the branch of knowledge dealing with these 道德规范的；伦理的

hail /heɪl/   *v.* ① ~ **sb/sth (as) sth** to describe sb/sth as being very good or special, especially in newspapers, etc. 赞扬（或称颂）……为…… ② to signal to a taxi or a bus, in order to get the driver to stop 招手（请出租车或公共汽车停下）

n. [U] small balls of ice that fall like rain 雹；冰雹

transparent /trænsˈpærənt/   *a.* ① (of a material or article) allowing light to pass through so that objects behind can be distinctly seen （材料，物品）透明的 ② (of an organization or its activities) open to public scrutiny（机构或其活动）受公众监督的；透明的

unionize /ˈjuːnɪənaɪz/   *v.* become or cause to become members of a trade union （使）加入工会；（使）成立工会

purportedly /pəˈpɔːtɪdli/   *ad.* believed or reputed to be the case 据称

virtuous /ˈvɜːtʃuəs/   *a.* having or showing high moral standards 道德高尚的；有德性的

shortfall /ˈʃɔːtfɔːl/   *n.* a deficit of sth required or expected 赤字；不足

offset /ˈɒfset/   *n.* [C] a consideration or amount that diminishes or balances the effect of a contrary one 弥补；抵消

	v. counteract sth by having an equal and opposite force or effect 抵消；弥补
authenticate /ɔːˈθentɪkeɪt/	*v.* prove or show sth, especially a claim or an artistic work to be true or genuine 证明……是确实的；证明……是真的
fake /feɪk/	*n.* ① a thing that is not genuine; a forgery or sham 假货；赝品 ② a person who appears or claims to be sth that they are not 冒充者；骗子 *a.* not genuine; counterfeit 伪造的；假冒的
dispose of	throw away sth that you no longer want or need 丢掉；处理
flood with	fill or suffuse completely 把……装满（或充满）
at intervals	① with time between, not continuously 不时 ② with spaces between 相隔一定的距离
refer to	① mention or allude to 提及；谈到 ② describe or denote; have as a referent 指；表示 ③ read or otherwise use (a source of information) in order to ascertain sth; consult 查阅；参考
keep pace with	move, increase, change, etc. at the same speed as sb/sth 与……步调一致；与……并驾齐驱
roll out	introduce or start to use a new product 发布；推出（新产品）
associate with	connect sb or sth with sth else in one's mind 把（某人或物）与其他人（或物）联系在一起
according to	as reported or stated by or in 根据，按照；据……所述
attribute to	regard sth as being caused by 把……归因于；把……归咎于
responsible for	being the primary cause of sth and able to be blamed or credited for it 负责的；承担（谴责或表扬）的

▪ Reading Comprehension

Understanding the text

Answer the following questions.

1. How many times does a person wear each garment before discarding it?

    A. Five.

    B. Six.

    C. Seven.

    D. Eight.

2. What changes have taken place in the clothing industry as the result of the rise of fast fashion?

    A. People bought five times fewer pieces of clothing.

    B. People kept clothing for far longer.

    C. People will not buy a new fast fashion brand dress each week without thinking twice.

    D. The market is filled with cheap, poorly-made garments.

3. Which of the following is not true about the way to keep pace with the fast changing consumer tastes?

    A. By releasing two or more collections at certain times each year.

    B. By selling clothes at a low price.

    C. By constantly pushing out new product.

    D. By producing clothing at high speed.

4. In what way do fast fashion's low price points rely on even lower producing costs?

    A. Not to exploit workers.

    B. To create acceptable working conditions for workers.

    C. To outsource production to companies all over the world in search of cheap labor.

    D. To pay employees far beyond the minimum wage.

5. According to a recent investigation, how much were workers in Los Angeles being paid?

    A. As little as $4.3 an hour.

    B. As little as $2.77 an hour.

C. As much as $11 an hour.

D. As much as $27.7 an hour.

6. Which of the following is the right way to avoid supporting fast fashion for consumers?

A. To buy from more ethical brands.

B. To purchase brand new clothing.

C. Not to keep something longer in your closet.

D. To buy clothes at a low price.

Critical thinking

Work in pairs and discuss the following questions.

1. Why is fast fashion also called disposable fashion?

2. How does fast fashion affect consumers?

3. What are the environmental effects of fast fashion?

▄ Language Enhancement

Words in use

Fill in the blanks with the words given below. Change the form when necessary. Each word can be used only once.

impulse	disastrous	release	accelerate	sponsor
intersperse	exploit	collapse	reveal	claim

1. Police have _____ no further details about the accident.

2. The company plans to _____ television programs as part of its marketing strategy.

3. A few pretty songs are _____ between tragic stories to supply comic relief.

4. Chemical leaks have had a (an) _____ effect on wildlife.

5. All of a sudden, a section of the railway bridge _____ in the severe earthquake.

6. Eric had an almost irresistible _____ to hug the beautiful girl in front of him.

7. The Ferrari Mondial can _____ from 0 to 60 mph in 6.3 seconds.

8. If no one _____ the items, they will become Crown property.

9. It is _____ in the latest newspaper that Frank and his children had received death threats after being kidnapped.

10. What should the young people in the workplace do to prevent their employers from _____ them?

Expressions in use

Fill in the blanks with the expressions given below. Change the form when necessary. Each expression can be used only once.

dispose of	refer to	flood with	at intervals
keep pace with	roll out	attribute to	responsible for

1. On other systems the time may only change _____ of 10 or 20 milliseconds, and the accuracy may vary from one change to the next.

2. Governments around the world increasingly _____ artificial intelligence to help promote economic growth.

3. When I give my presentation, I will probably have to _____ my notes.

4. She walked quickly, and her decisiveness helped me to _____ her.

5. He _____ the firm's success _____ the efforts of the managing director.

6. If you are barbecuing with coal, be sure you know how to properly _____ them.

7. Parking lot owners should be legally _____ protecting vehicles.

8. If your home doesn't have enough sun light, get more lamps and _____ the place _____ brightness.

Sentence structure

I. Complete the following sentences by translating the Chinese into English, using "the more ... the more ... " structure.

Model: The longer the war lasts, _____

_____. (那里的人们受难就越多。)

→The longer the war lasts, the more the people there will suffer.

1. The more difficult the questions are, _____

_____（他回答出来的可能性就越小）.

2. The harder you work, _____

_____（进步就越大）.

3. The earlier you start, _____

_____（回来得就越早）.

II. Rewrite the following sentences by using "whatever/however/wherever".

Model: No matter how knowledgeable the professor is, he is not able to teach his students everything they want to know.

→However knowledgeable the professor is, he is not able to teach his students everything they want to know.

1. No matter where they went, they were closely followed by security guards.

_____

_____.

2. No matter how carefully I explained, she still didn't understand.

_____

_____.

3. Don't lose heart no matter what difficulties you meet.

_____

_____.

Extensive Reading

Fast Fashion: Its Detrimental Effect on the Environment

1   "Fast fashion" clothing retailers make cheap and fashionable clothing,

but the cost comes at a high price. According to the UN Environment Programme, the fashion industry is the second-biggest consumer of water and is responsible for 8–10% of global carbon emissions—more than all international flights and maritime shipping combined.

What Is Fast Fashion?

2   The term "fast fashion" has become more prominent in conversations surrounding fashion, sustainability and environmental consciousness. The term refers to "cheaply produced and priced garments that copy the latest catwalk styles and get pumped quickly through stores in order to maximise on current trends". The fast fashion model is called so because it involves rapid design, production, distribution and marketing, which means that retailers are able to pull large quantities of greater product variety and allow consumers to get more fashion and product differentiation at a low price.

Why Is Fast Fashion Bad?

3   According to Business Insider, fashion production comprises 10% of total global carbon emissions, as much as the European Union. It dries up water sources and pollutes rivers and streams, while 85% of all textiles go to dumps each year. Even washing clothes releases 500, 000 tons of microfibres into the ocean each year, the equivalent of 50 billion plastic bottles.

4   The Quantis International 2018 report found that the three main drivers of the industry's global pollution impacts are dyeing and finishing, yarn preparation and fibre production. The report also established that fibre production has the largest impact on freshwater withdrawal and ecosystem quality due to cotton cultivation, while the dyeing and finishing, yarn preparation and fibre production stages have the highest impacts on resource depletion, due to the energy-intensive processes based on fossil fuel energy.

5   According to the UN Framework Convention on Climate Change, emissions from textile manufacturing alone are projected to skyrocket by 60% by 2030. The time it takes for a product to go through the supply

chain, from design to purchase, is called a "lead time". This results in the fashion industry producing obscene amounts of waste.

The Environmental Impacts of Fast Fashion

Water

6   Among the environmental impacts of fast fashion are the depletion of non-renewable sources, emission of greenhouse gases and the use of massive amounts of water and energy. The fashion industry is the second largest consumer industry of water, requiring about 700 gallons to produce one cotton shirt and 2,000 gallons of water to produce a pair of jeans. Business Insider also cautions that textile dyeing is the world's second-largest polluter of water, since the water leftover from the dyeing process is often dumped into ditches, streams or rivers.

Microplastics

7   Furthermore, brands use synthetic fibres like polyester, nylon and acrylic which take hundreds of years to biodegrade. A 2017 report from the International Union for Conservation of Nature (IUCN) estimated that 35% of all microplastics, tiny pieces of non-biodegradable plastic, in the ocean come from the laundering of synthetic textiles like polyester.

8   According to the documentary released in 2015, *The True Cost*, the world consumes around 80 billion new pieces of clothing every year, 400% more than the consumption twenty years ago. The average American now generates 82 pounds of textile waste each year. The production of leather requires large amounts of feed, land, water and fossil fuels to raise livestock, while the tanning process is among the most toxic in all of the fashion supply chain because the chemicals used to tan leather, including mineral salts, formaldehyde, coal-tar derivatives and various oils and dyes, is not biodegradable and contaminates water sources.

Energy

9   The production of making plastic fibres into textiles is an energy-

intensive process that requires large amounts of petroleum and releases volatile particulate matter and acids like hydrogen chloride. Additionally, cotton, which is in a large amount of fast fashion products, is also not environmentally friendly to manufacture. Pesticides deemed necessary for the growth of cotton presents health risks to farmers. To counter this waste caused by fast fashion, more sustainable fabrics that can be used in clothing include wild silk, organic cotton, linen, hemp and lyocell.

"Slow Fashion"

10 Slow fashion is the widespread reaction to fast fashion, the argument for hitting the brakes on excessive production, overcomplicated supply chains, and mindless consumption. It advocates for manufacturing that respects people, the environment and animals.

11 The World Resources Institute suggests that companies need to design, test and invest in business models that reuse clothes and maximise their useful life. The UN has launched the Alliance for Sustainable Fashion to address the damages caused by fast fashion. It is seeking to "halt the environmentally and socially destructive practices of fashion".

12 One way that shoppers are reducing their consumption of fast fashion is by buying from secondhand sellers like Thred Up Inc. and Poshmark, both based in California, USA; shoppers send their unwanted clothes to these websites and people buy those clothes at a lower price than the original. Another solution is renting clothes, like the US-based Rent the Runway and Gwynnie Bee, the UK based Girl Meets Dress, and the Dutch firm Mud Jeans that leases organic jeans which can be kept, swapped or returned.

13 Governments need to be more actively involved in the fashion industry's damaging effects. UK ministers rejected a report by members of parliament to address the environmental effects of fast fashion. On the other hand, French president, Emmanuel Macron has made a pact with 150 brands to make the fashion industry more sustainable.The best advice on reducing fast fashion comes from Patsy Perry, senior lecturer in fashion marketing at the University of Manchester, who says, "Less is always more."

Unit 7

Retail of Textile and Clothing

High and low isn't such a novelty thing. It's how young people interpret the life we've been given. It's how we look at luxury brands. It's how we look at heritage brands.

—*Virgil Abloh*

International retailers of clothing are believed to be the key drivers of the globalization of the clothing industry.

—*Gary Gereffi*

Pre-Reading Activities

1. Listen to the recording and answer the following questions.

(1) What is retailing?

(2) According to McKinsey & Company's "Perspectives on retail and consumer goods", what should successful retail companies do?

2. Listen to the recording again and fill in the blanks based on what you hear.

Once the clothes have been designed and manufactured, they need to be sold. Retailers make 1. _____ purchases for resale three to six months before the consumers are able to buy the clothes in-store. Retail comes in a 2. _____ of shapes, sizes and platforms, and involves the sale of goods or 3. _____ directly to consumers. The most commonly known retail stores, often 4. _____ as "brick and mortar" operations, include 5. _____ and specialty department stores, discount and chain stores, consumer electronics, home improvement and office supply retailers, sporting goods, footwear and jewelry stores, as well as pharmacies, food stores, 6. _____ and so on. The US retail landscape is changing rapidly as technology disrupts many traditional retail companies. Traditional store closings have 7. _____ store openings in the US in recent years, while 8. _____ continues to grow.

3. Discuss the following questions with your partner.

(1) Do you know any famous brick-and-mortar stores? Which store do you like best? Why?

(2) Do you think it is necessary for college students to buy luxury products? Why or why not?

International Trade in Garments

1   **Garments** have been perhaps the most international consumer products dating back to the late 1800s when Paris **couturiers**—led by the House of Worth—**dictated** the styles that **affluent** women around the world wore. Those garments were composed of rich, imported fabrics and **trimmings**: silks from China, woolens and **velvets** from England, **damasks** and **lace** from Italy. Back then, **representatives** from the first department stores in the United States, Marshall Field in Chicago, R. H. Macy in New York, and John Wanamaker in Philadelphia, voyaged to Paris in order to purchase haute couture samples, which their workrooms **translated into** styles suitable for America's consumers.

2   Clothing **licensing** leads to globalization. In the 1980s, the fashion **boom popularized** designer jeans and athletic shoes and accelerated the globalization of fashion, as licensing became a lifeline to the fashion business, especially couture fashion houses. In fashion licensing, a design house collects a royalty payment between 3 to 10 percent of wholesale volume, from an outside manufacturer who produces and **markets** the merchandise. Licensing **enabled** designers to put their **trademarks** on clothes, handbags, jewelry, shoes, and perfume quickly and painlessly. Licensing turned designers like Pierre Cardin and Calvin Klein into household names as they built billion dollar empires marketing sofas, bedsheets, clocks, and even frying pans to an international **marketplace**.

3   Modern fashion houses have no national **borders**. For example, an **outfit** that carries an Italian fashion label might have well been created in Milan, by a team of British, French and American designers, and **manufactured** by **contractors** from countries in Vietnam, Korea and Mexico.

4   International trade has **affected** countries all over the world. This shift toward globalization is evident in trade **statistics**. In 1999, the four leading exporters of clothing were China, Italy, the United States and Germany. The five largest importers of clothing were the United States, Germany, Japan, the United Kingdom and France. France, despite the continuing **prominence** of Paris in the world of fashion design, had only $5.69 billion in clothing **exports** in 1999.

5   Fashion marketers have **been forced to** think globally in order to cater to a cross section of international shoppers whom they now serve directly. No longer do consumers have to travel abroad to find the top brands, as a lot of brands have **blanketed** the world with **boutiques** from Buenos Aires to Tokyo.

6   Media, technology and international consumers are all crucial factors in international trade. Thanks to the Internet, Hollywood and cable television, there is not much difference between consumers in Spain and those in the United States, who are all **exposed to** the same trends, **celebrity** role models, and popular music **simultaneously**. Furthermore, globalization has leveled the playing field, enabling retailers to **compete** in the multibillion-dollar fashion game, as these **discount chains** have learned to master the mechanics of delivering fast fashion **at rock bottom prices**.

7   Fashion entered a new era in the 1990s, as the world's **buoyant** high-tech industries broke the pattern for formal dress codes. Jeans, khakis, and knitwear replaced suits as the new **corporate** uniform. To seek higher ground, the high fashion industry could no longer **bank on** dress-up clothes. Marketers thus found a new **hook** to **captivate** consumers: **accessories** like handbags, shoes and watches, which could be **plastered with showy** designer logos and **coordinated with** casual clothes. Furthermore, accessories delivered higher profit **margins** than apparel, making them even more attractive to fashion marketers.

8   Luxury accessories thus became the **focal** point of such European

conglomerates. By owning a roster of fashion brands, some clothing giants have benefited from economies of scale. The luxury fashion boom began in the late 1990s as women began collecting trinkets, the $1,000 baguette handbags, for instance, that they wore to dress up their casual clothes. The accessories boom underscored how international fashion trends have become as shoppers in Tokyo, Paris, and New York all flocked to buy the hot designer handbags of the season.

9　　　Industry experts agree the next frontier is the Internet, which takes globalization to new heights. With the click of a mouse from their laptops, the world's consumers can conveniently shop the shelves of Harrods in London, L.L. Bean in Maine, Neiman Marcus in Texas, as well as Ebay, the auction website that features fashion merchandise offered by millions of individuals from around the world.

10　　　More than anything, the Internet has exposed the world's consumers to an infinite range of choices at all price ranges. Apparel retailers will be led to provide an access to goods and discounts in 2020s, no matter which medium consumers choose to purchase, such as smartphones, desktops or the brick-and-mortar stores, multiple channels can bring the same benefits to shopping experience. It is the ultimate example of globalization, built on the back of rapidly changing media, which has further democratized fashion and reduced the world into a single and accessible marketplace.

Notes

The House of Worth is a French house of high fashion that specializes in haute couture, ready-to-wear clothes, and perfumes. The historic house was founded in 1858 by English-born designer Charles Frederick Worth. It continued to operate under his descendants until 1952 and closed in 1956. The House of Worth fashion brand was revived in 1999.

Marshall Field's was a department store in Chicago, Illinois, founded in the 19th century that grew to become a large chain before being acquired by Macy's, Inc. in 2005.

R. H. Macy's is a department store which sells apparel, housewares, jewelry and beauty supplies. Some locations may also sell furniture.

John Wanamaker Department Store was one of the first department stores in the United States. Founded by John Wanamaker in Philadelphia, it was influential in the development of the retail industry including as the first store to use price tags.

L. L. Bean is an American, privately held retail company founded in 1912 by Leon Leonwood Bean. The company is headquartered where it was founded, in Freeport, Maine. It specializes in clothing and outdoor recreation equipment.

royalty payment　版权费

Buenos Aires　布宜诺斯艾利斯（阿根廷首都）

khakis　卡其裤

Harrods　哈罗德百货，位于伦敦

New words and phrases

garment /ˈgɑːmənt/	*n.* a piece of clothing （一件）衣服
couturier /kuˈtjʊərieɪ/	*n.* a person who designs, makes and sells expensive, fashionable clothes, especially for women （尤指）女装设计师；时装裁缝；女装商人
dictate /dɪkˈteɪt/	*v.* ① to control or influence how sth happens 支配；摆布；决定 ② to say words for sb else to write down 口述 ③ to tell sb what to do, especially in an annoying way（尤指以令人不快的方式）指使；强行规定
	n. [usually pl.] (*formal*) an order or a rule that you must obey 命令；规定
affluent /ˈæfluənt/	*a.* having a lot of money and a good standard of living 富裕的

trimming /ˈtrɪmɪŋ/	*n.* ① [C, usually pl.] material that is used to decorate sth, for example along its edges 装饰材料；镶边饰物 ② [pl.] the extra things that it is traditional to have for a special meal or occasion（菜肴的）配料；额外的事物 [pl.] the small pieces of sth that are left when you have cut sth 修剪下来的东西；剪屑
velvet /ˈvelvɪt/	*n.* a type of cloth made from silk, cotton or nylon, with a thick soft surface 丝绒；立绒；经绒；天鹅绒
damask /ˈdæməsk/	*n.* a type of thick cloth, usually made from silk or linen, with a pattern that is visible on both sides 花缎；锦缎
lace /leɪs/	*n.* ① [U] a delicate material made from threads of cotton, silk, etc. that are twisted into a pattern of holes 网眼织物；花边；蕾丝 ② [C]= shoelace 鞋带
	v. ① to be fastened with laces; to fasten sth with laces 由带子系紧；把……用带子系牢 ② to put a lace through the holes in a shoe, a boot, etc. 给（鞋、靴等）穿鞋带 ③ to twist sth together with another thing 使编织（或交织、缠绕）在一起
representative /ˌreprɪˈzentətɪv/	*n.* ① a person who has been chosen to speak or vote for sb else or on behalf of a group 代表 ② Representative (*abbr.* Rep.) (in the US) a member of the House of Representatives, the Lower House of Congress; a member of the House of Representatives in the lower house of a state parliament （美国）众议院议员
	a. typical of a particular group of people 典型的；有代表性的

licensing /ˈlaɪsnsɪŋ/	*n.* the act of giving a formal (usually written) authorization 授权
boom /buːm/	*n.* ① a sudden increase in trade and economic activity; a period of wealth and success（贸易和经济活动的）激增；繁荣 ② a period when sth such as a sport or a type of music suddenly becomes very popular and successful（某种体育运动、音乐等）突然风靡的时期
	v. ① to make a loud deep sound 轰鸣；轰响 ② to say sth in a loud deep voice 以低沉有力的声音说 ③ to have a period of rapid growth; to become bigger, more successful, etc. 迅速发展；激增；繁荣昌盛
popularize /ˈpɒpjələraɪz/	*v.* ① to make a difficult subject easier to understand for ordinary people 使通俗化；使普及 ② to make a lot of people know about sth and enjoy it 宣传；宣扬；推广
market /ˈmɑːkɪt/	*v.* to advertise and offer a product for sale; to present sth in a particular way and make people want to buy it 推销；促销
enable /ɪˈneɪbl/	*v.* ① to make it possible for sb to do sth 使能够；使有机会 ② to make it possible for sth to happen or exist by creating the necessary conditions 使成为可能；使可行；使实现
trademark /ˈtreɪdmɑːk/	*n.* ① a name, symbol or design that a company uses for its products and that cannot be used by anyone else 商标 ② a special way of behaving or dressing that is typical of sb and that makes them easily recognized（人的行为或衣着的）特征；标记
marketplace /ˈmɑːkɪtpleɪs/	*n.* ① [C] an open area in a town where a market is held 集市；市场 ② [sing.] the activity of competing with other companies to buy and sell goods, services, etc. 市场竞争

border /ˈbɔ:də(r)/

n. ① the line that divides two countries or areas; the land near this line 国界；边界；边疆；边界地区 ② a strip around the edge of sth such as a picture or a piece of cloth 镶边；包边

v. ① to share a border with another country or area 和……毗邻；与……接壤 ② to form a line along or around the edge of sth 沿……的边；环绕……；给……镶边

outfit /ˈaʊtfɪt/

n. ① [C] a set of clothes that you wear together, especially for a particular occasion or purpose 全套服装；装束（尤指为某场合或目的）② a group of people working together as an organization, business, team, etc. 团队；小组；分队 ③ a set of equipment that you need for a particular purpose 全套装备；成套工具

v. to provide sb/sth with equipment or clothes for a special purpose 装备；配置设备；供给服装

manufacture /ˌmænjuˈfæktʃə(r)/ *v.* ① to make goods in large quantities, using machinery （用机器）大量生产，成批制造 ② to produce a substance 生成，产生（一种物质）

contractor /kənˈtræktə(r)/

n. a person or company that has a contract to do work or provide goods or services for another company 承包人；承包商；承包公司

affect /əˈfekt/

v. ① to produce a change in sb/sth 影响 ② to attack sb or a part of the body; to make sb become ill/sick 侵袭；使感染 ③ to make sb have strong feelings of sadness, pity, etc. （感情上）深深打动；使悲伤（或怜悯等）

statistics /stəˈtɪstɪks/

n. a branch of applied mathematics concerned with the collection and interpretation of quantitative data and the use of probability theory to estimate population parameters 统计学；统计

prominence /ˈprɒmɪnəns/	*n.* the state of being important, well known or noticeable 重要；突出；卓越；出名
export /ɪkˈspɔːt/	*n.* ① [U] the selling and transporting of goods to another country 出口；输出 ② [C, usually pl.] a product that is sold to another country 出口产品；输出品
	v. ① to sell and send goods to another country 出口；输出 ② to introduce an idea or activity to another country or area 传播，输出（思想或活动）③ (*computing* 计) to send data to another program, changing its form so that the other program can read it 输出；移出；调出
blanket /ˈblæŋkɪt/	*v.* to cover sth completely with a thick layer 覆盖
	n. ① a large cover, often made of wool, used especially on beds to keep people warm 毯子；毛毯 ② [usually sing.] ~ **of sth** a thick layer or covering of sth 厚层；厚的覆盖层
	a. [only before noun] including or affecting all possible cases, situations or people 包括所有情形（或人员）的；总括的；综合的
boutique /buːˈtiːk/	*n.* a small shop/store that sells fashionable clothes or expensive gifts 时装店；精品店；礼品店
expose /ɪkˈspəʊz/	*v.* ① to let sb find out about sth by giving them experience of it or showing them what it is like 使接触；使体验 ② to show sth that is usually hidden 暴露；显露；露出
celebrity /səˈlebrəti/	*n.* ① [C] a famous person 名人；名流 ② [U] the state of being famous 名望；名誉；著名
simultaneously /ˌsɪmlˈteɪniəsli/	*ad.* at the same instant 同时地

compete /kəmˈpiːt/   *v.* ① ~ **(with/against sb) (for sth)** to try to be more successful or better than sb else who is trying to do the same as you 竞争；对抗 ② ~ **(in sth) (against sb)** to take part in a contest or game 参加比赛（或竞赛）

discount /ˈdɪskaʊnt/   *n.* an amount of money that is taken off the usual cost of sth 折扣

v. ~ **sth (as sth)** (*formal*) to think or say that sth is not important or not true 认为……不重要；对……不全信；低估

chain /tʃeɪn/   *n.* ① a group of shops/stores or hotels owned by the same company 连锁商店 ② a series of connected metal rings, used for pulling or fastening things; a length of chain used for a particular purpose 链子；链条；锁链 ③ a series of connected things or people 一系列，一连串（人或事）

buoyant /ˈbɔɪənt/   *a.* ① (of prices, business activity, etc. 价格、商业活动等) tending to increase or stay at a high level, usually showing financial success 看涨的；保持高价的；繁荣的 ② cheerful and feeling sure that things will be successful 愉快而充满信心的；乐观的 ③ floating, able to float or able to keep things floating 漂浮的；能够漂起的；有浮力的

corporate /ˈkɔːpərət/   *a.* ① connected with a corporation 公司的 ② forming a corporation 组成公司（或团体）的；法人的 ③ involving or shared by all the members of a group 社团的；全体的；共同的

hook /hʊk/   *n.* ① a thing designed to catch people's attention 吸引人之物；噱头 ② a curved piece of metal, plastic or wire for hanging things on, catching fish with, etc. 钩；钓钩；挂钩；鱼钩

captivate /ˈkæptɪveɪt/	*v.* to keep sb's attention by being interesting, attractive, etc. 迷住；使着迷
accessory /əkˈsesəri/	*n.* ① [usually pl.] a thing that you can wear or carry that matches your clothes, for example a belt or a bag （衣服的）配饰 ② [usually pl.] an extra piece of equipment that is useful but not essential or that can be added to sth else as a decoration 附件；配件；附属物
showy /ˈʃəʊi/	*a.* so brightly coloured, large or exaggerated that it attracts a lot of attention 显眼的；艳丽的；花哨的
margin /ˈmɑːdʒɪn/	*n.* ① (*business* 商) = profit margin 利润 ② the empty space at the side of a written or printed page 页边空白；白边
focal /ˈfəʊkl/	[only before noun] central; very important; connected with or providing a focus 中心的；很重要的；焦点的；有焦点的
conglomerate /kənˈglɒmərət/	*n.* ① (*business* 商) a large company formed by joining together different firms 联合大公司；企业集团 ② [sing.] (*formal*) a number of things or parts that are put together to form a whole 合成物；组合物；聚合物
trinket /ˈtrɪŋkɪt/	*n.* a piece of jewellery or small decorative object that is not worth much money （价值不高的）小首饰；小装饰物
underscored /ˌʌndəˈskɔː(r)/	*v.* = underline 划线
flock /flɒk/	*v.* to go or gather together somewhere in large numbers 群集；聚集；蜂拥 *n.* ① a group of sheep, goats or birds of the same type （羊或鸟）群 ② a large group of people, especially of the same type （尤指同类人的）一大群

frontier /ˈfrʌntɪə(r)/	*n.* ① the limit of sth, especially the limit of what is known about a particular subject or activity（学科或活动的）尖端，边缘 ② (*BrE*) a line that separates two countries, etc.; the land near this line 国界；边界；边境
infinite /ˈɪnfɪnət/	① very great; impossible to measure 极大的；无法衡量的 ② without limits; without end 无限的；无穷尽的
translated into	change from one form or medium into another 转化为
be forced to	to make sb accept sth that they do not want 被迫
expose to	to let sb find out about sth or experience sth 使接触；使体验
at rock bottom prices	lowest possible price 最低价
bank on	expect it to happen and rely on it happening 指望
plaster with	to cover ever inch of some surface with sth 涂上
coordinate with	bring (components or parts) into proper or desirable coordination correlation 搭配
a roster of	a list of 一系列的
dress up	make sth appear superficially attractive 打扮
flock to	move as a crowd or in a group 蜂拥；聚集

▪ Reading Comprehension

Understanding the text

Answer the following questions.

1. According to the passage, what are the most international consumer products in the late 1800s?

2. What made designer jeans and sneakers popular?

3. Why did designers like Pierre Cardin and Calvin Klein become household names?

4. How do you understand the sentence "Modern fashion houses have no national borders" in paragraph 3?

5. Please explain the sentence "International trade has affected countries all over the world" in your own words.

6. What are the important factors in international trade?

7. Why can discount chains compete in fashion games?

8. Why did the fashion industry enter a new era in the 1990s?

9. According to paragraph 8, why are accessories popular?

10. What will happen to apparel retailers in 2020s?

Critical thinking

I. Make a Presentation.

If you are a brand retailer, what factors do you think will affect clothing retail? Try to prepare for a presentation in class.

II. Work in pairs and discuss the following questions.

1. How do you understand the sentence "The apparel retail is the most competitive businesses in the textile industry"?

2. According to this article, are there any new trends in the global clothing trade? What are they?

◼ Language Enhancement

Words in use

Fill in the blanks with the words given below. Change the form when necessary. Each word can be used only once.

dictate	representative	affluent	enable	showy
evident	prominence	infinite	expose	compete

1. Smoking used to be more common among _____ people.

2. Since he was color-blind, he cannot distinguish _____ flowers.

3. The spacecraft will _____ scientists to study the solar system more.

4. Humanistic courses should be gaining _____ in college curricula.

5. After a long period of full preparation, it is _____ that she is sure to win the contest.

6. What gives him the right to _____ to us what we should say?

7. The company is ready to _____ in international trade.

8. We want to _____ college students to as much traditional literature and culture as possible.

9. With _____ care, Mary told him the truth.

10. She seemed to take it for granted that she should give a speech as a _____.

Banked cloze

Fill in the blanks by selecting suitable words from the word bank. You may not use any of the words more than once.

A. exposure	F. retailer	K. crucial
B. source	G. allows	L. associate
C. extravagant	H. representative	M. apparel
D. provide	I. replicate	N. rent
E. crack	J. convenient	O. acquire

Entrepreneur defines a trade show as "an exhibition for companies in a specific industry to showcase and demonstrate their new products and services." This 1. _____ goes on to say that most trade shows aren't open to the public, and only company 2. _____ and press organizations are given access. Fashion trade shows are special events where fashion designers and brand owners show off their new fashions to potential clients and 3. _____, and these events are held all over the world.

In most cases, the facilitators of fashion trade shows 4. _____ out huge exhibition halls where designers show off all the fashions they've been working on during the previous year. Location is 5. _____ in these types of trade shows, and designers fight tooth and nail for the booth locations that will 6. _____ them with the best 7. _____. While the best fashion trade shows are in fashion capitals like New York City and Paris, pretty much every major city has one of these events at least once per year, which means you'll have the opportunity to show off your fashions without having to travel a great distance.

A fashion trade show 8. _____ you to select business partners and gain fame in the fashion industry at one 9. _____ location. Retailers rely on these trade shows for 10. _____ sourcing, so whether you focus on contemporary apparel, children's apparel, or any other type of clothing, a trade fair is one of the first places you should visit as you make a name for yourself as a designer.

Expressions in use

Fill in the blanks with the expressions given below. Change the form when necessary. Each expression can be used only once.

date back	translated into	cater to	bank on
coordinate with	dress up	flock to	be composed of

1. In the United States, rapidly changing styles _____ a desire for novelty and individualism.

2. Or you could _____ the future potential of this project.

3. Your work will _____ those three items.

4. I believe all the hard work will _____ profits.

5. The magazine is not a new one. It _____ to the 1930s at least.

6. Children like to _____ as doctors.

7. People _____ the mall to buy discounted goods.

8. Jeans color should _____ the color of your shoes.

■ Translation

I. Translate the following paragraph into Chinese.

Marketing operates at both the wholesale and retail levels. Companies that do not sell their own products at retail must place those products at wholesale prices in the hands of retailers, such as boutiques, department stores, and online sales companies. They use fashion shows, catalogs, and a sales force armed with sample products to find a close fit between the manufacturer's products and the retailer's customers. Marketers for companies that do sell their own products at retail are primarily concerned with matching products to their own customer base. At both the wholesale and the retail level, marketing also involves promotional activities such as print and other media advertising aimed at establishing brand recognition and brand reputation for diverse characteristics such as quality, low price, or trendiness.

By the early 21st century, the Internet had become an increasingly important retail outlet, creating new challenges (e.g., the inability for customers to try on clothes prior to purchase, the need for facilities designed to handle clothing returns and exchanges) and opening up new opportunities for merchandisers (e.g., the ability to provide customers with shopping opportunities 24 hours per day, affording access to rural customers). In an era of increasingly diverse shopping options for retail customers and of intense price competition among retailers, merchandising has emerged as one of the cornerstones of the modern fashion industry.

II. Translate the following paragraph into English.

随着消费者数据变得更加容易获得,许多新的机会(和失误)为想要挖掘购物者数据的零售商们打开了大门。排在首位的是个性化,它是一种多用途的工具。对于营销者来说,这意味着能够更好地定位销售方案和广告。对于零售商来说,这可能意味着一切,从为定制商店提供精心策划的产品推荐到为特定人群进行更大的品牌推广。对于购物者来说,这是专门为他们量身定做一个产品的感觉。购物者和零售商之间的关系变得更像朋友之间的对话。2020 年,随着疫情席卷零售业,围绕个性化的讨论略有改变。然而,在一些公司，个性化一如既往地走在了前列。一些零售商继续扩大商店概念，依靠个性化来推动体验。

Paragraph Writing

How to write a cause-and-effect essay

In this unit, you'll learn how to write a cause-and-effect essay. Each day you face situations that require cause-and-effect analysis. Some are daily events; others mark important life decisions. When you analyze causes, you ask the question why. If you think about questions, you will probably come up with one or more reasons to explain a certain behavior. In this case, you have analyzed the causes. Similarly, when you analyze effects, you ask what it is. You usually consider what the possible results or consequences of some action could be. By doing this, you have explained the effects. Very often, causes have a single effect while multiple effects may be the result of one single cause. Look at the examples below:

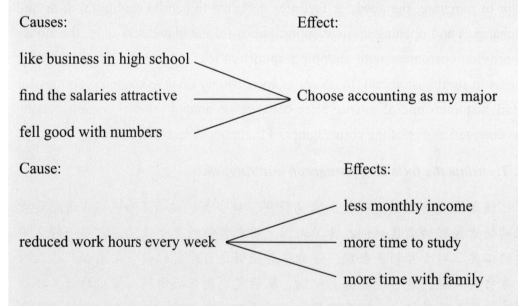

Causes: Effect:

like business in high school

find the salaries attractive   Choose accounting as my major

fell good with numbers

Cause: Effects:

                                           less monthly income

reduced work hours every week   more time to study

                                           more time with family

Since an essay of 150 words is hard to include both causes and effects, you are required to focus on either causes or effects instead of both. The usual logical patterns we can follow is like this:

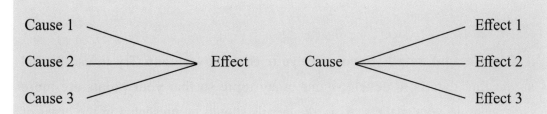

When writing such a kind of essay, it is important to follow these steps:

1. Have a clear understanding of causes and effects. Causes should answer the question: why did something happen? The effects must explain: what happened because of it? Look at the following two examples:

(1) There are several reasons why living in dorms makes attending college somewhat difficult.

* less sleep time
* more temptations to skip class
* more distractions from studying

(2) Attending college has somehow changed my personality in positive ways.

* more confident
* more knowledgeable
* more communicative

2. Write an effective thesis statement. Make sure you clearly discuss causes or effects in the thesis statement. You're encouraged to list the causes or effects in your thesis statement so that your readers can easily predict what you will discuss in the essay. Look at the examples:

(1) Movie stars lead very stressful lives because they have too little privacy, too much pressure, and too little time for themselves.

(2) There are three major components to blame for our societal bad behavior: longer working hours, too much time on technology, and the tension caused by anger.

3. Use necessary cause effect transitions. Include appropriate transitions between paragraphs in your essay to make your ideas flow more smoothly. Transitions indicating causes include: *since, the first reason is, because, another reason is. Transitions indicating effects include: result in, another effect is, consequently,* *therefore.*

4. Give factual details to support your thesis statement. Try to use factual supporting details to develop your essay topic so that your points are more convincing to your readers. Also, the details should be presented in the order of importance: arrange the information from least to most important or vice versa.

5. Conclude the essay by summarizing the causes or effects. Give a brief summary of the causes or effects in your conclusion, followed by your final evaluation or comments, whether positive or negative. They can be an opinion, a prediction, a warning, a suggestion, or an appeal.

What's more, try to remember the following logical requirements:

(1) Do not oversimplify causes.

(2) Beware especially of making a mistake in logic. For instance, if a disease broke out soon after X became Minister of Public Health, we would probably make a mistake in logic in saying that the epidemic broke out because X was made minister. Remember sequence does not always indicate causation.

(3) Distinguish between direct and indirect causes and between major and minor causes.

(4) Do not omit links in a chain of causes.

(5) Be objective and support the analysis with solid, factual evidence.

Now, let's look at an example and see how to write a cause-and-effect essay.

In recent years more and more foreigners are beginning to have an interest in the Chinese language. According to a survey, Chinese-language courses are given in more than 3,000 schools in over 109 foreign countries.

There are various factors for the widespread interest in Chinese. One of the most common causes is that China's ancient culture has attracted many people in the world. More and more foreigners are becoming interested in this oriental country and enjoying its splendid culture. To understand and study this culture, one must learn the Chinese language. Another contributing cause is that China is enjoying rapid economic growth, and its economic and cultural exchange with other countries has increased continuously. Consequently, people who can speak Chinese will be popular in the job market. Perhaps the primary factor is that the Chinese language itself has charm in its characters and pronunciation: The picturesque Chinese characters arouse in some foreigners a sense of beauty and are considered attractive. The pronunciation of Putonghua is so beautiful and musical that it has attracted many foreigners to learn.

Many people believe that the Chinese language will become a more important language in the world in decades to come. As China will open further to the outside world, its language is sure to spread more widely in the world.

Topic		Why learn the Chinese language?
Introduction	Thesis statement	In recent years more and more foreigners are beginning to take an interest in the Chinese language.
Body	Cause 1	China's ancient culture has attracted many people in the world.
	Cause 2	China is enjoying rapid economic growth, and its economic and cultural exchange with other countries has increased continuously.
	Cause 3	The Chinese language itself has charm in its characters and pronunciation.
Conclusion	Effect	The Chinese language is sure to spread more widely in the world.

Paragraph 1 is the introduction of the essay, serving as thesis statement why more and more foreigners are beginning to take an interest in the Chinese language. In paragraph 2, the author provides three reasons why they learn the Chinese language. The last paragraph is the conclusion, in which the results caused by the reasons are summarized.

Structured writing

Read the sample essay and see how the effects are developed.

Topic

Career opportunities created by the aging population

Introduction:

Thesis statement: The aging population in America creates career opportunities.

Body:

Effect 1: Medical and health professions, as well as other professions, are needed.

Effect 2: Lawyers are in need.

Effect 3: Businessmen see huge opportunities.

Conclusion:

The expansion of America's elderly population will provide good job opportunities in many areas.

Sample essay

Old is suddenly in. By 2050, 25 percent of all Americans will be older than 65, up from 14 percent in 1995. The change poses profound questions for the government and society, of course. But it also creates career opportunities in the medical and health professions, in law, and in business.

Medical and health professions, as well as other professions, are imperatively needed. "In addition to physicians, we're going to need more sociologists, biologists, and urban planners," says professor Edward Schneider of the University of Southern California's School of Gerontology.

Lawyers can specialize in "elder law", which covers everything from trusts and estate planning to nursing-home abuse and age discrimination.

Businessmen see huge opportunities in the elder market because the baby boomers, 74 million strong, are likely to be the wealthiest group of retirees in human history. "Any student who combines an expert knowledge in gerontology with, say, an MBA or law degree will have a license to print money," one professor says. Margarite Santos is a 21-year-old senior at USC. She began college as a biology major but found she was "really bored with bacteria". So she took a class in gerontology and discovered that she liked it. She says, "I did volunteer work in retirement homes and it was very satisfying."

In short, the expansion of America's elderly population will provide good job opportunities in many areas.

Write an essay of no less than 150 words on the topic "What causes people's irrational behavior". You can follow the outline given below.

Topic:

What causes people's irrational behavior

Introduction:

Thesis statement: Irrational behavior is caused by various factors but two stand out.

Body:

Cause 1: Uncontrolled emotions

Cause 2: Stressed feeling

Conclusion:

Uncontrolled emotions and too much stress can result in irrational behavior.

More topics:

Topic 1: Pressure on students

Topic 2: The effects of smartphone addiction

The Fashion Retailer Retail and Culture

1   When I was 16 years old, I didn't know what to study at university. I spent some time soul-searching about what subject I liked at school and why. I liked economics especially its social aspect. I have always been **passionate** about history and tried to put all the pieces of puzzle together even there is no single **version** of the truth. So I finally decided to study sociology (the study of social life, social change, social causes and **consequences** of human behavior).

2   While studying sociology, I realized how different sociology, psychology or economics, to mention a few **disciplines**, approached the same **issue**. There was a lack of any **multidisciplinary collaboration**. In my opinion, every science must connect to each other while **breaking down silos** and **biases**. Similar issues arise when individually analyzing business results from a financing, marketing, operations, planning or designing **perspectives**.

3   I became passionate about retail for the same reasons as I'm interested in the **humanities**. Retail is a **reflection** of society. Understanding retail is knowing social, political and economical trends. A retail business should be **analysed** from different **angles** like social **phenomena**. As customers, we have been focusing too much on the product or the **front-end** side of the business, that is, too much attention to customer-facing side of a company. Therefore, "Brand-Consumer" is a **superficial** relationship. Equally, brands have been focusing too much on the product and its financial **ratios**. In *No Logo: Taking Aim at the Brand Bullies*, Klein argued that **globalization** was a **process whereby** corporations discovered that **profits** didn't lie in production (**outsourced** to low-wage workers in developing countries), but

in creating branded identities people adopt in their lifestyles. Few people cared about the fact that brands were "cool" and supply chains were dark.

4   Retail has changed. B2C businesses **communicate with** consumers through selling not only products but experiences involving the feeling of **community**. This is why some famous brands are not selling directly on Amazon but are **centralizing** its **distribution** network. **Thanks to** the Internet, including sites and social media, B2B players are implementing the "new business model" to **get closer to** their customers. On one hand, consumers **care about** retailers' ethics (e.g. social and environmental impacts); on the other hand, brands are **participating in** the political **arena** and taking a stand in social and environmental issues. This e-business pattern is not an exception, but the rule.

5   The 20th century saw the emergence of brands and logos. Early versions of logos were developed by ancient Egyptians. In the Middle Ages, shops and pubs used **signage** to **represent** what they did. The Bass Brewery's label **incorporating** its triangle logo for **ale** was the first trademark to be **registered** under the Trade Mark Registration Act 1875. Since then, the power of brands grew **exponentially**. Coke, for example, changed the appearance of Santa Claus.

6   Brands are niche "**flexible**" **religions**. When shopping for clothes, or any other product (i.e. food, cars, smart phones, watches), consumers **fulfill** their needs. Their **motivation** goes from physiological (i.e. clothes can **insulate** against cold or hot conditions) to **self-actualization** needs (i.e. lifestyle, **status**), as described by Abraham Maslow and **commented** on the fashion **pyramid**. Customers are demanding more **transparency** and originality. Values become essential elements of the story. Humans are political and philosophical animals and many have the chance to satisfy needs **other than** simply physiological ones. These are, for instance, **universal** values such as freedom, equality or **solidarity**.

7   Brands are positioned from mass-market segmentation to supreme. **At the**

top of the pyramid, brands tend to underline creativity, **craftsmanship**, experience and art. Brands are somehow niche religions with their own values, lifestyle, language and **symbolism**. **Even if** it's not 100% **accurate**, if you tell me what you buy (category of products, segment, style, size, colors, brand, etc.), I'll tell you who you are. Or like Joshua Becker said on Internet, "what we consume determines the lives we live". These are some interesting news items I read in relation to brand's activism or brands "investing" on "sustainable back-end".

8   A coffee shop is investing $100m in small businesses and communities. According to CNBC, it is creating a new $100 million fund aimed at investing in community development projects and small businesses in areas **populated** by people of color. The plan is to invest the funds over the next four years in 12 US cities with populations that are black, indigenous or people of color, including Atlanta, Detroit, Los Angeles, Philadelphia, and Houston. Small businesses and neighborhood development projects are the target **beneficiaries** for this program.

9   A company's latest stores format includes Rise and Unite. Combine serves as a community **centerpiece** connected by sport. The products are **reflective** of what the community is interested in: locally **curated**, every-day essentials at the best price **matched with** the newness of select seasonal **offerings**. Additionally, the company hires people who live in the local community.

10   Producer and **rapper** Pharrell Williams is **coupling** his entrepreneurial spirit and activism with the launch of Black Ambition, an **incubator** for Black and Latinos entrepreneurs launching **startups** in tech, design, healthcare and consumer products. The initiative has raised an undisclosed amount of funds from Chanel, the Chan Zuckerberg Initiative, the Rockefeller Foundation and the Visa Foundation among others and is seeking cooperation with more companies for **mentorship**.

11   As customers are more **sensitive** about the **back-end** side of business,

brands are becoming active organizations **in defense of** "their" values (e.g. human rights, **sustainability**). Being a cool brand is not only a marketing thing but also involving other departments or perspectives. It is very important to know about the retail culture and brand ethics to help us have detailed information about their procedure according to how things go in this very industry.

Notes

Middle Ages in European history was the period of history between classical antiquity and the Italian Renaissance.

The Bass Brewery was founded in 1777 by William Bass in Burton upon Trent, England. The main brand was Bass Pale Ale, once the best selling beer in the UK.

The Trade Marks Registration Act The passing of the 1875 Trade Marks Registration Act created a system that, for the first time, allowed UK businesses to formally register and gain legal protection for their trademarks, preventing other companies from copying their product identity. The Trade Marks Registration Act and the rules thereunder, with introduction, notes, and practical directions as to registering trade marks, together with the Merchandise Marks Act, 1862, with notes and a copious index to the whole.

B2C 企业对消费者的电子商务模式（Business-to-Customer）

Amazon 美国亚马逊公司

B2B 企业间电子商务（Business-to-Business）

Santa Claus 圣诞老人

CNBC 美国全国广播公司财经频道（Consumer News and Business Channel）

Rockefeller Foundation 洛克菲勒基金会

Visa Foundation Visa 基金会

New words and phrases

passionate /ˈpæʃənət/ — *a.* having or showing strong feelings of enthusiasm for sth or belief in sth 热诚的；狂热的

version /ˈvɜːʃn/ — *n.* ① a form of sth that is slightly different from an earlier form or from other forms of the same thing 变体；变种 ② a description of an event from the position of a particular person or group of people （从不同角度的）说法；描述 ③ a film/movie, play, piece of music, etc. that is based on a particular piece of work but is in a different form, style or language （电影、剧本、乐曲等的）版本；改编形式；改写本

consequence /ˈkɒnsɪkwəns/ — *n.* ① [C] ~ **(for sb/sth)** a result of sth that has happened 结果；后果 ② [U] (*formal*) importance 重要性

discipline /ˈdɪsəplɪn/ — *n.* ① [U] the practice of training people to obey rules and orders and punishing them if they do not; the controlled behaviour or situation that results from this training 训练；训导；纪律；风纪 ② [C] a method of training your mind or body or of controlling your behaviour; an area of activity where this is necessary 训练方法；行为准则；符合准则的行为 ③ [U] the ability to control your behaviour or the way you live, work, etc. 自制力；遵守纪律 ④ [C] an area of knowledge; a subject that people study or are taught, especially in a university 知识领域；（尤指大学的）学科；科目

v. ① ~ **sb (for sth)** to punish sb for sth they have done 惩罚；处罚 ② ~ **yourself** to control the way you behave and make yourself do things that you believe you should do 自我控制；严格要求（自己）③ to train sb, especially a child, to obey particular rules and control the way they behave 训练；训导；管教

issue /ˈɪʃuː/

n. ① [C] a problem or worry that sb has with sth （有关某事的）问题；担忧 ② [C] an important topic that people are discussing or arguing about 重要议题；争论的问题 ③ [U] the act of supplying or making available things for people to buy or use 发行；分发

v. ① ~ **sth (to sb)** to make sth known formally 宣布；公布；发出 ② [often passive] ~ **sth (to sb)** | ~ **sb with sth** to give sth to sb, especially officially （正式）发给；供给 ③ to produce sth such as a magazine, article, etc. 出版；发表

multidisciplinary
/ˌmʌltidɪsəˈplɪnəri/

a. involving several different subjects of study（涉及）多门学科的

collaboration /kəˌlæbəˈreɪʃn/

n. ① [U,C] ~ **(with sb) (on sth)** | ~ **(between A and B)** the act of working with another person or group of people to create or produce sth 合作；协作 ② [C] a piece of work produced by two or more people or groups of people working together 合作成果（或作品）③ [U] (*disapproving*) the act of helping the enemy during a war when they have taken control of your country 通敌；勾结敌人

silo /ˈsaɪləʊ/　　　　　　*n.* ① a tall tower on a farm used for storing grain, etc. 筒仓 ② an underground place where nuclear weapons or dangerous substances are kept（核武器的）发射井；（危险物品的）地下贮藏库 ③ an underground place where silage is made and stored 青贮窖

bias /ˈbaɪəs/　　　　　　*n.* ① a strong feeling in favour of or against one group of people, or one side in an argument, often not based on fair judgement 偏见；偏心；偏向 ② [usually sing.] an interest in one thing more than others; a special ability 偏爱；特殊能力

v. ~ **sb/sth** (towards/against/in favour of sb/sth) to unfairly influence sb's opinions or decisions 使有偏见；使偏心；使偏向

perspective /pəˈspektɪv/　　*n.* ① [C] ~ **(on sth)** a particular attitude towards sth; a way of thinking about sth 态度；观点；思考方法 ② [C] a view, especially one in which you can see far into the distance 景观；远景 ③ [U] the ability to think about problems and decisions in a reasonable way without exaggerating their importance 客观判断力；权衡轻重的能力

humanity /hjuːˈmænəti/　　*n.* ① **the humanities** [pl.] the subjects of study that are concerned with the way people think and behave, for example literature, language, history and philosophy 人文学科 ② [U] people in general（统称）人；人类 ③ [U] the state of being a person rather than a god, an animal or a machine 人性 ④ [U] the quality of being kind to people and animals by making sure that they do not suffer more than is necessary; the quality of being humane 人道；仁慈

reflection /rɪˈflekʃn/   *n.* (*AmE* reflexion) ① [C] an image in a mirror, on a shiny surface, on water, etc. 映像；映照出的影像 ② [U] the action or process of sending back light, heat, sound, etc. from a surface（声、光、热等的）反射 ③ [C] a sign that shows the state or nature of sth 反映；显示；表达 ④ [U] careful thought about sth, sometimes over a long period of time 沉思；深思；审慎的思考

analyse /ˈænəlaɪz/   *v.* (*AmE* analyze) to examine the nature or structure of sth, especially by separating it into its parts, in order to understand or explain it 分析

angle /ˈæŋgl/   *n.* ① a particular way of presenting or thinking about a situation, problem, etc. 观点；立场；角度 ② a position from which you look at sth 角度 ③ the space between two lines or surfaces that join, measured in degrees 角 ④ the direction that sth is leaning or pointing in when it is not in a vertical or horizontal line 斜角；角度

v. ① to move or place sth so that it is not straight or not directly facing sb/sth 斜移；斜置 ② to present information, a report, etc. based on a particular way of thinking or for a particular audience 从（某角度）报道；以（某观点）提供信息

phenomenon /fəˈnɒmɪnən/   *n.* ① (pl. phenomena) a fact or an event in nature or society, especially one that is not fully understood 现象 ② (pl. phenomenous in *NAmE*) a person or thing that is very successful or impressive 杰出的人；非凡的人（或事物）

front-end /ˈfrʌnt end/   *a.* the side that is forward or prominent 前端的

superficial /ˌsuːpəˈfɪʃl/   *a.* ① (often *disapproving*) not studying or looking at sth thoroughly; seeing only what is obvious 粗略的；肤浅的；粗枝大叶的；浅薄的 ②appearing to be true, real or important until you look at it more carefully 表面的；外面的；外表的 ③ (of a wound or damage 伤口或损坏) only affecting the surface and therefore not serious 表层的；表皮的 ④ (*disapproving*) not concerned with anything serious or important and lacking any depth of understanding or feeling 浅薄的；肤浅的

ratio /ˈreɪʃiəʊ/   *n.* ~ **(of A to B)** the relationship between two groups of people or things that is represented by two numbers showing how much larger one group is than the other 比率；比例

globalization /ˌɡləʊbəlaɪˈzeɪʃn/   *n.* [U] the fact that different cultures and economic systems around the world are becoming connected and similar to each other 全球化；全世界化(世界各地的文化和经济体系日益关联)

process /ˈprəʊses/   *n.* ① a series of things that are done in order to achieve a particular result（为达到某一目标的）过程；进程 ② a series of things that happen, especially ones that result in natural changes（事物发展，尤指自然变化的）过程，步骤，流程 ③ a method of doing or making sth, especially one that is used in industry 做事方法；工艺流程；工序

v. ① to treat raw material, food, etc. in order to change it, preserve it, etc. 加工；处理 ② to deal officially with a document, request, etc. 审阅；审核；处理（文件、请求等）

whereby /weəˈbaɪ/   *ad.* (*formal*) by which; because of which 凭此；借以；由于

profit /ˈprɒfɪt/   *n.* ① [C, U] ~ **on sth** ~ **from sth** the money that you make in business or by selling things, especially after paying the costs involved 利润；收益；赢利 ② [U] (*formal*) the advantage that you get from doing sth 好处；利益；裨益

v. ~ **(by/from sth)** (*formal*) to get sth useful from a situation; to be useful to sb or give them an advantage 获益；得到好处；对……有用（或有益）

outsource /ˈaʊtsɔːs/   *v.* to arrange for sb outside a company to do work or provide goods for that company 交外办理；外包

community /kəˈmjuːnəti/   *n.* ① [sing.] all the people who live in a particular area, country, etc. when talked about as a group 社区；社会 ② [C+sing./pl.v.] a group of people who share the same religion, race, job, etc. 团体；社团；界 ③ [C] (*biology* 生) a group of plants and animals growing or living in the same place or environment（动植物的）群落

centralize /ˈsentrəlaɪz/   *v.* to give the control of a country or an organization to a group of people in one particular place 集权控制；实行集中

distribution /ˌdɪstrɪˈbjuːʃn/   *n.* ① [U, C] the way that sth is shared or exists over a particular area or among a particular group of people 分配；分布 ② [U] the act of giving or delivering sth to a number of people 分发；分送 ③ [U] the system of transporting and delivering goods（商品）运销；经销；分销

arena /əˈriːnə/   *n.* ① (*formal*) an area of activity that concerns the public, especially one where there is a lot of opposition between different groups or countries 斗争场所；竞争舞台；活动场所 ② a place with a flat open area in the middle and seats around it where people can watch sports and entertainment 圆形运动场；圆形剧场

signage /ˈsaɪnɪdʒ/   *n.* signs collectively (especially commercial signs or posters) 招牌；标识

represent /ˌreprɪˈzent/   *v.* ① [often passive] to be a member of a group of people and act or speak on their behalf at an event, a meeting, etc. 代表 ② to act or speak officially for sb and defend their interests 作为……的代言人；维护……的利益 ③ (not used in the progressive tenses 不用于进行时) to be sth 等于；相当于；意味着 ④ [no passive] to be an example or expression of sth 成为……实例；成为典型；体现 ⑤ (not used in the progressive tenses 不用于进行时) to be a symbol of sth 作为……的象征；象征；代表

incorporate /ɪnˈkɔːpəreɪt/   *v.* ① ~ **sth (in/into/within sth)** to include sth so that it forms a part of sth 将……包括在内；包含；吸收；使并入 ② (*business* 商) [often passive] to create a legally recognized company 注册成立

ale /eɪl/   *n.* ① [U, C] a type of beer, usually sold in bottles or cans. 麦芽啤酒 ② [U] (*old-fashioned*) beer generally（泛指）啤酒

register /ˈredʒɪstə(r)/   *v.* ① ~ **(at/for/with sth)** | ~ **sth (in sth)** | ~ **(sb) as sth** | to record your/sb's/sth's name on an official list 登记；注册 ② (often used in negative sentences 常用于否定句) to notice sth and remember it; to be noticed 注意到；记住；受到注意 ③ [usually passive] to send sth by mail, paying extra money to protect it against loss or damage 把……挂号邮寄 ④ (*formal*) to make your opinion known officially or publicly （正式地或公开地）发表意见；提出主张

exponentially /ˌekspəuˈnenʃəli/   *ad.* in an exponential manner 以指数方式地

flexible /ˈfleksəbl/   *a.* ① (*approving*) able to change to suit new conditions or situations 能适应新情况的；灵活的；可变动的 ② able to bend easily without breaking 柔韧的；可弯曲的；有弹性的

religion /rɪˈlɪdʒən/   *n.* ① the belief in the existence of a god or gods, and the activities that are connected with the worship of them 宗教信仰 ② a particular interest or influence that is very important in your life 特别的兴趣；重大的影响 ③ [C] one of the systems of faith that are based on the belief in the existence of a particular god or gods 宗教；教派

fulfill /fʊlˈfɪl/   *v.* ① to make sb feel happy and satisfied with what they are doing or have done 使高兴；使满意 ② to do or have what is required or necessary 履行；执行；符合；具备 ③ to do or achieve what was hoped for or expected 实现

motivation /ˌməʊtɪˈveɪʃn/   *n.* the psychological feature that arouses an organism to action toward a desired goal; the reason for the action; that which gives purpose and direction to behavior 动机

insulate /ˈɪnsjuleɪt/   *v.* to protect sth with a material that prevents heat, sound, electricity, etc. from passing through 使隔热；使隔音；使绝缘

self-actualization /ˌself ˌæktʃuəlaɪˈzeɪʃn/   *n.* [U] the fact of using your skills and abilities and achieving as much as you can possibly achieve 自我实现（利用自身技能取得尽可能大的成就）

status /ˈsteɪtəs/   *n.* ① [U] [C, usually sing.] the social or professional position of sb/sth in relation to others 地位；身份；职位 ② [U] the situation at a particular time during a process （进展的）状况；情形

comment /ˈkɒment/   *v.* ~ **(on/upon sth)** to express an opinion about sth 表达意见

n. ① [C, U] ~ **(about/on sth)** sth that you say or write which gives an opinion on or explains sb/sth 议论；评论；解释 ② [sing., U] criticism that shows the faults of sth 批评；指责

pyramid /ˈpɪrəmɪd/   *n.* ① an organization or a system in which there are fewer people at each level as you get near the top 金字塔式的组织（或系统）② a large building with a square or triangular base and sloping sides that meet in a point at the top （古埃及的）金字塔 ③ (*geometry* 几何) a solid shape with a square or triangular base and sloping sides that meet in a point at the top 锥体；棱锥体

transparency /trænsˈpærənsi/   *n.* ① [U] the quality of sth, such as an excuse or a lie, that allows sb to see the truth easily 显而易见；一目了然 ② [U] the quality of sth, such as glass, that allows you to see through it 透明；透明性 ③ (also *informal* **tranny**) [C] a picture printed on a piece of film, usually in a frame, that can be shown on a screen by shining light through the film 幻灯片；透明正片

universal /ˌjuːnɪˈvɜːsl/   *a.* ① true or right at all times and in all places 普遍存在的；广泛适用的 ② done by or involving all the people in the world or in a particular group 普遍的；全体的；全世界的；共同的

solidarity /ˌsɒlɪˈdærəti/   *n.* ~ **(with sb)** support by one person or group of people for another because they share feelings, opinions, aims, etc. 团结；齐心协力；同心同德；相互支持

craftsmanship /ˈkrɑːftsmənʃɪp/   *n.* ① the level of skill shown by sb in making sth beautiful with their hands 手艺；技艺 ② the quality of design and work shown by sth that has been made by hand 精工细作

symbolism /ˈsɪmbəlɪzəm/   *n.* [U] the use of symbols to represent ideas, especially in art and literature （尤指文艺中的）象征主义；象征手法

accurate /ˈækjərət/   *a.* ① correct and true in every detail 正确无误的 ② able to give completely correct information or to do sth in an exact way 精确的；准确的 ③ an accurate throw, shot, weapon, etc. hits or reaches the thing that it was aimed at 准确的（掷、射、击等）

populate /ˈpɒpjuleɪt/   v. ① [often passive] to live in an area and form its population inhabit 居住于；生活于；构成……的人口 ② to move people or animals to an area to live there 迁移；移居；殖民于 ③ (*computing* 计) to add data to a document （给文件）增添数据；输入数据

beneficiary /ˌbenɪˈfɪʃəri/   n. ① a person who gains as a result of sth 受益者；受惠人 ② a person who receives money or property when sb dies 遗产继承人

centerpiece /ˈsentəpiːs/   n. ① the central or most important feature 中心 ② something placed at the center of something else (as on a table) 放在（桌子）中央的摆饰

reflective /rɪˈflektɪv/   a. ① (*formal*) thinking deeply about things 沉思的；深思的 ② surfaces send back light or heat （指物体表面）反射热的，反光的 ③ **~ of sth** typical of a particular situation or thing; showing the state or nature of sth 典型的；代表性的；体现状态（或本质）的

curate /ˈkjʊərət/   v. to hold or plan the display of (a collection or exhibit) 操持（收藏品或展品）的展出

offering /ˈɒfərɪŋ/   n. ① sth that is produced for other people to use, watch, enjoy, etc. 用品；剧作；作品；供消遣的产品 ② sth that is given to a god as part of religious worship 祭品；供品

rapper /ˈræpə(r)/   n. a person who speaks the words of a rap song 说唱歌手

couple /ˈkʌpl/   v. ① [usually passive] **~ A (to B) | ~ A and B together** to join together two parts of sth, for example two vehicles or pieces of equipment （把车辆或设备等）连接；结合

n. ① [C+sing./pl.v.] two people who are seen together, especially if they are married or in a romantic or sexual relationship（人）一对；（尤指）夫妻；情侣 ② [sing.+sing./pl.v.] ~ **(of sth)** two people or things 两人；两件事物 ③ ~ **(of sth)** a small number of people or things 几个人；几件事物

incubator /ˈɪŋkjubeɪtə(r)/　*n.* ① a piece of equipment in a hospital which new babies are placed in when they are weak or born too early, in order to help them survive（体弱或早产婴儿）恒温箱 ② a machine like a box where eggs are kept warm until the young birds are born 孵化器

startup /stɑːtʌp/　*n.* ① the act of setting in operation 启动 ② the act of starting a new operation or practice 创业公司

mentorship /ˈmentəʃɪp/　*n.* 导师制；辅导教师；师徒制

sensitive /ˈsensətɪv/　*a.* ① ~ **(about/to sth)** easily offended or upset 易生气的；易被惹恼的；神经过敏的 ② ~ **(to sth)** aware of and able to understand other people and their feelings 体贴的；体恤的；善解人意的 ③ able to understand art, music and literature and to express yourself through them 感觉敏锐的；艺术感觉好的；有悟性的 ④ ~ **(to sth)** reacting quickly or more than usual to sth 敏感的；过敏的 ⑤ ~ **(to sth)** able to measure very small changes 灵敏的

back-end /ˌbæk ˈend/	*a.* [only before noun] ① relating to the end of a period or process 结束的；结尾的 ② (*computing* 计) (of a device or program 设备或程序) not used directly by a user, but used by a program or computer 后端的；后台的（指不归用户直接使用，而由程序或电脑使用）
sustainability /səˌsteɪnəˈbɪləti/	*n.* the property of being sustainable 可持续性
breaking down	① the act of disrupting an established order so it fails to continue 打破；分解开 ② a mental or physical breakdown 崩溃
communicate with	to link，to talk with 与某人联系；交流；与……沟通
thanks to	just that, owing to 由于；幸亏
get closer to	make the distance or difference between two people or things smaller 接近
care about	think of; concern about/for/over 担心；关心
participate in	take part in; go through 参加；参与
other than	different from, different than 除了；不同于
at the top of	在顶部；在最高地位
even if	even though 即使；虽然
matched with	make and fit 使和……相匹配
in defense of	plead, guard 为……辩护；保卫

▪ Reading Comprehension

Understanding the text

Answer the following questions.

1. What subject does the author finally decide to study?

   A. Sociology.

B. Economics.

C. History.

D. Psychology.

2. According to the passage, what do you know about retail?

A. Retail is a reflection of society.

B. A retail business should be analysed like social phenomena: from different angles.

C. B2C businesses communicate with consumers through selling not only products but experiences.

D. All of the above.

3. When consumers are shopping for clothes or any other product, _____.

A. clothes can insulate against cold or hot conditions

B. their motivation goes from physiological to self-actualization needs

C. they just fulfill their physiological needs

D. they buy what they want

4. The early versions of logos were developed by _____.

A. ancient Egyptians

B. officers in the Middle Ages

C. Bass Brewery

D. Coke

5. At the bottom, brands emphasize _____.

A. category of products, segment, style, size, colors

B. values, lifestyle, language, symbolism

C. creativity, craftsmanship, experience and art

D. all of the above

6. Pharrell Williams is coupling his entrepreneurial spirit and activism with the launch of Black Ambition. The initiative has raised an undisclosed amount of funds from _____.

A. Chanel

B. the Chan Zuckerberg Initiative

C. the Rockefeller Foundation and the Visa Foundation

D. all of the above

7. From the passage, we know _____.

    A. it is very important to know about the retail culture

    B. brand ethics can help us have detailed information about their procedure according to how things go in this very industry

    C. retail is very important to enterprises

    D. both A and B

Critical thinking

Work in pairs and discuss the following questions.

1. Do you think brand is important? Why, or why not?

2. When purchasing, what factors do you think are important?

■ Language Enhancement

Words in use

Fill in the blanks with the words given below. Change the form when necessary. Each word can be used only once.

passionate	accurate	approach	perspective	sensitive
superficial	process	consequence	universal	motivation

1. I firmly believe that the newspaper is much more _____ than television news.

2. It is the _____ truth that the more focused the child is, the easier it is for him to succeed in the future.

3. We can tell the characteristics of painters: they are _____, impulsive and unique.

4. I think such a view is a _____ analysis based on the information supplied by the mass media.

5. Iceberg melting is one of _____ of global warming.

6. We have many communicative _____ to talk with foreigners.

7. The learning _____ is long but satisfying.

8. I think I have a very different _____ when it comes to the relationship between money and happiness.

9. Employment care is a politically _____ issue.

10. Most people said that pay was their main _____ for working.

Expressions in use

Fill in the blanks with the expressions given below. Change the form when necessary. Each expression can be used only once.

communicate with	break down	at the top of	get closer to
other than	in defense of	aim at	match with

1. Tom never speaks to his mother _____ to ask for something.

2. Bees _____ each other by dancing.

3. I wrote down what I wanted to say to him _____ the page.

4. Students should _____ contributing something useful to the class rather than damaging the interests of the class.

5. It's because of the practicalities of education that you have to _____ the curriculum into specialist subjects.

6. The advertisement they designed doesn't _____ our value system.

7. Obviously, as we _____ success, the truth gradually emerges.

8. We should show our respect for heroes who died or injured _____ the country in the World War II.

Sentence structure

I. Complete the following sentences by translating the Chinese into English, using "no matter ..." structure.

Model: Apparel retailers will be led to provide an access to goods and discounts in 2020s, _____,
(无论消费者选择通过哪种媒介购买) such as smartphones, desktops or the brick-and-mortar stores, multiple channel can bring

the same benefits to shopping experience.

→Apparel retailers will be led to provide an access to goods and discounts in 2020s, <u>no matter which medium consumers choose to purchase</u>, such as smartphones, desktops or the brick-and-mortar stores, multiple channel can bring the same benefits to shopping experience.

1. _____ (无论你喜欢哪一条裙子), you can have a try.

2. _____ (无论你说什么), I won't believe you.

3. _____ (无论你什么时候有空), you can come here and talk with me.

4. _____ (不论我去哪儿), I will never forget my hometown.

5. _____ (不论他多么努力), he finds it difficult to fulfill it.

II. Rewrite the following sentences by using "When doing ..." structure.

Model: When consumers are shopping for clothes, or any other product, they fulfill their needs.

→When shopping for clothes, or any other product, consumers fulfill their needs.

1. When you complete your assignments, you should be careful.

_____

_____.

2. When she sings in public, she always feels nervous.

_____

_____.

3. When she goes shopping, she always buys things at a discount.

_____

_____.

Is the Fashion Retail Industry the Right Choice for You?

Fashion—what is the first thing that comes into your mind at the mention of this word? Glamour? Celebrities? Fame? Well, the fashion industry is all of that and more. According to Fashion United, its market value is about 385.7 billion dollars, which makes up 4% of the market share. Many industries together form the fashion industry as we know it today, such as merchandising, textile and designing. A major section of fashion careers is the retail business.

If you want to make it big in the fashion retail industry, this insightful article will answer all the questions you might have.

What is fashion retailing?

Fashion retailing is the section of business that acts as an intermediary between the manufacturers and customers. It can be defined as the process of "buying clothes from the manufacturer and selling them to the customers".

How to get into fashion retail?

Fashion retail is a diverse sector of the fashion industry, where you will need a combination of a strong academic background and interpersonal skills to do well in the profession. The following are some steps that will guide you to a career in fashion retailing.

Step 1: Educational qualifications

In order to enter this industry, you must have a bachelor's degree in fashion retailing, fashion marketing, fashion merchandising or a related field. A master's degree is preferred as it increases your chances of getting a job.

Step 2: Interpersonal skills

· COMMUNICATION SKILLS: First and foremost, you must have effective communications skills. This skill is essential for any career you pursue. Since communication is a two-way street, you must also have good listening skills. In this field, you will be required to interact with manufacturers and designers on a regular basis. Other than this, you will also need to understand the customers' demands and feedback.

· PRODUCT KNOWLEDGE: It is important that you are aware of the business you are in and understand it fully. You keep up to date with the products your organisation sells, the market trends, popular products and merchandising techniques that can benefit your company.

· POSITIVE OUTLOOK: Since fashion is a relatively competitive industry, it is vital that you mould yourself according to the business. You must always have a positive outlook, irrespective of the challenges that come your way. You must also be flexible and be aware that the retail industry is unconventional in many aspects, such as working hours.

· PEOPLE SKILLS: Since a major part of your job will involve people, you must have the capability of building effective professional relationships. It adds value to your résumé if you are a team player and possess team building qualities.

· SELLING SKILLS: The most important skill that a professional in fashion retail should have is the ability to pull off the sale of products. You must have persuasion skills, ability to explain the product features, intricate product knowledge and fashion marketing skills.

Step 3: Internship

Pursuing an internship along with your course can help you learn a lot about the trade. Fashion internships prove to be extremely beneficial if you want to get some on-the-job training before you get to the real deal.

Reasons to work in fashion retail

· HEALTHY COMPETITION: Fashion is a highly competitive industry, especially with the constantly changing trends and new styles coming in. If you are game for being nurtured in a competitive environment, this is the career for you. You will be constantly working towards becoming better at your job and getting paid for it too!

· WORK WITH PEOPLE: Working with people can be exciting and terrifying at the same time. However, if you are a "people person," then this job is for you. From interacting with customers to collaborating with designers, you will get to communicate with people from all walks of life and make a living out of it.

· ENTHUSIASM: For fashion enthusiasts, this is the best place to be. You will be surrounded by everything fashionable—apparels, accessories, footwear, etc. Moreover, you will get the chance to work with the who's who in the industry.

· GOOD PAY: As advertised on payscale.com, the average salary of a fashion retailer is about £36,937 per year.

Fashion Retailer Job Profile

While there are many positions you can assume after a degree in fashion retail and each job comes with a different set of responsibilities, the basic job role of a fashion retailer includes:

· identifying the right suppliers for your organization;
· devising strategies for sales and product development;
· reviewing old products and selecting new ones;
· fashion marketing;
· ensuring the smooth transition of products, from the manufacturers to the consumers;
· setting sales targets and motivating subordinates to achieve them;
· budgeting and promotion of products;
· maintaining effective professional relationships with designers and manufacturers;
· maintaining effective customer relationships and working on feedback received from them;
· attending trade fairs and conferences;
· implementing changes in strategies based on customer feedback.

Job Roles in Fashion Retailing

· FASHION RETAIL BUYER: The job role of a fashion buyer requires them to curate the sale range that a clothing line will put up on sale. The success quotient of professionals who assume this role relies on the depth of their industry knowledge. If you take up this role, you will be given a huge set of responsibilities that will influence the decisions of the brand.

· VISUAL MERCHANDISER: As a visual merchandiser, your job responsibilities will include boosting sales by promoting merchandise, make the product on display visually appealing so that it piques customers' interests and improve the shopping experience of customers who walk in to the store.

· STORE MANAGER: Store managers are responsible for all the in-store functions like generating profits, dealing with customer grievances, ensuring a smooth shopping experience for customers and supervising the staff. They usually work in close proximity with visual merchandisers and make the store visually appealing so that customers are attracted towards the merchandise on display.

· SALES ASSISTANT: If you assume the role of a sales assistant, you will be required to assist shoppers and make their shopping experience easy and trouble-free. Your responsibilities would also include dealing with customer complaints and helping them with information about products.

If you are interested in becoming a fashion retailer, you should learn about key concepts of management such as business strategy, operations and project management, leadership management, theoretical knowledge and their practical applications. You could acquire leadership and decision-making skills that are crucial in the business setting today.

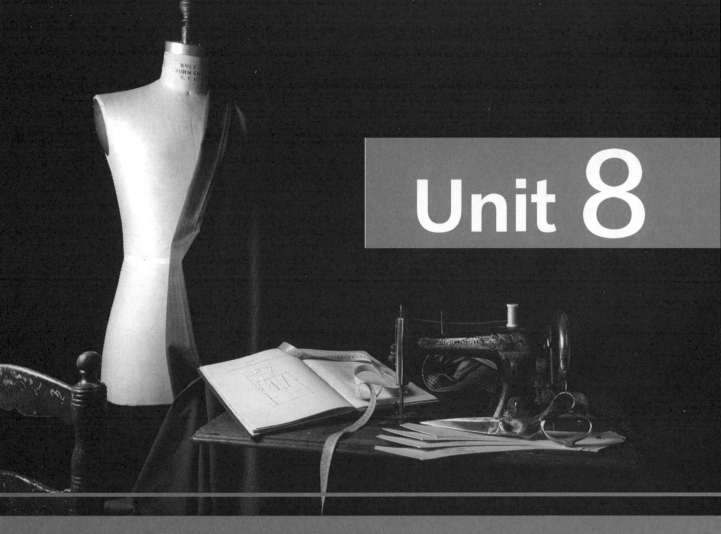

Unit 8

Changes in Fashion and Clothing

Fashion is very important. It is life-enhancing, like everything that gives pleasure, and it is worth doing well.

—*Vivienne Westwood*

Fashion is part of the daily air and it changes all the time, with all the events. You can even see the approaching of a revolution in clothes. You can see and feel everything in clothes.

—*Diana Vreeland*

Pre-Reading Activities

1. Listen to the recording and answer the following questions.

(1) What is the role of social media platforms in fashion industry?

(2) How does social media affect fashion designers?

2. Listen to the recording again and fill in the blanks based on what you hear.

In fashion industry, social media platforms can be used to interact with the (1) _____, as a means of networking with others in the industry and building an online (2) _____. Social media can also influence designers in (3) _____ ways that many designers create their fashions. Designers and (4) _____ players in fashion industry find (5) _____ for their designs and creations in many places: from recent (6) _____ to visits fashion capitals in the world, but the Internet and social media (7) _____ have created an entirely new approach that is getting both (8) _____ from fashion community.

3. Discuss the following questions with your partner.

(1) Do you think it is necessary for the fashion industry to be updated with new technology? Why, or why not?

(2) From your point of view, what factors can affect fashion designers?

How Is CAD Impacting on the Fashion Industry?

1　Too often, we tend to think of CAD as having **primarily** industrial **applications**. However, it's important to remember that CAD refers to a broad range of design programs, **encompassing** a number of creative fields as well as the AEC sector. One such field is the fashion industry, which has **keenly** felt the impact of the rise of CAD over the last few decades.

2　Now, we're seeing the link between fashion and CAD growing stronger than ever. Knowledge of major CAD programs has become an essential tool for many fashion designers, and the use of new design and manufacturing technologies are beginning to **spread from** the catwalk **to** the high street. Here, we're examining how CAD has already **had a vital impact upon** the fashion industry, and taking a look at what's in store for the future.

3　The development of the modern fashion industry began in the 19th century with the founding of the first fashion house in Paris, the House of Worth. Since that time, fashion styles have changed dramatically with the advent of social changes and shifts in consumer tastes. Despite this, the means by which designers created their pieces remained the same, using a pencil and paper to design.

4　Now, however, CAD has begun to make **inroads** into the world of fashion. It remains true that many fashion designers still **sketch** first drafts of a new design by hand—something which is no longer common in the AEC sector, as well as some other branches of product design. However, CAD programs are now increasingly used in fashion to create sketches, **prototypes** and designs. Fashion designers use 2D CAD software packages to create **initial** concepts for their designs. Then, when they are ready to create the full piece, they employ 3D CAD software to visualize their designs on virtual models.

5 From haute couture to the mass market, every sector of the fashion industry has felt the impact of CAD. As such, knowledge of CAD has become increasingly important to help secure a job in fashion design. It's easy to see why the industry has embraced CAD. Using CAD software, designers can create new sketches more quickly and more precisely. They can also easily adapt a single design to varying materials and patterns, and build upon existing designs to create new pieces.

6 CAD also allows designers to test out variants of their designs using different fabrics and swatches. As opposed to traditional design methods, designers using CAD can view these designs in both 2D and 3D, and make alterations immediately. Not only does this save time and reduce errors, but it also cuts costs down by letting designers view virtual versions of finished products without needing to purchase materials. It means that designers can view the entire design history of their product. Thus they can work out where they went wrong, or create different products from one initial idea. In short, CAD allows fashion designers to reduce waste and save money to work conveniently, and shorten the time taken from initial concept to finished piece.

7 In the fast-moving world of the fashion industry, with seasonal trends and consumer demands to contend with, the ability to use CAD to modify and create garments gives designers an edge. CAD also enables designers to tailor and customize existing styles, and produce new clothing to meet customers' needs and desires. Additionally, the advent of cloud-based CAD software has enabled designers to collaborate easily from any location. CAD is also being used in more innovative ways to create entirely new and more customizable products. These advances have made the creation of new clothing simpler, more collaborative and more cost-effective. Meanwhile, the ability to visualize designs in 3D further helps designers create garments that fit well and are ready to wear.

8 In the world of haute couture, a number of designers are using 3D printers

to create truly innovative and unique pieces. One of the most notable fashion designers to adopt the technology is Iris van Herpen, who has employed 3D-printed pieces in several of her collections, beginning with 2010's Crystallization. Van Herpen began using 3D printing after being "struck by its complexity and detailing", and her striking and sometimes unsettling pieces have pushed the boundaries of both fashion and technology. But she's far from the only one embracing 3D printing.

9   The ability to use CAD applications makes it possible to create designs more precise and intricate than ever before. This has particularly impacted on the footwear industries. We're seeing 3D-printed footwear coming to the running track courtesy of some of the biggest names in sportswear. Back in November, we reported on Reebok's Liquid Factory innovation lab, which created shoes with a "three-dimensional fit". However, they're far from the only company who see the potential in 3D printing.

10   The future craft 3D shoe aims to create shoes tailored to each individual's foot, "setting the athlete up for the best running experience". Another company, meanwhile, has focused on the top end of athletics, creating the new product alongside Olympic gold medallist and confirming their comm target to 3D printing by partnering with HP. We've also seen Under Armour and New Balance launch their own 3D-printed shoes, showing the serious interest from the athletics sector in the technology.

11   Both *Teen Vogue* and *Business of Fashion* have listed 3D fashion engineering as the most important burgeoning careers within the industry. While CAD has already had a major impact on the fashion industry, there are still exciting developments on the horizon. Noted futurist Ray Kurzweil predicted that, within the next decade, 3D-printed clothing would become common. Whilst we may still be a while off printing clothes in our own homes, the technology is already having an impact in certain areas. It's clear that CAD and manufacturing will continue to change this diverse industry, pushing forward the boundaries of fashion.

Notes

AEC is used to describe software collections, sectors of business and a range of communities that rely on CAD. AEC stands for architecture, engineering and construction. There are many overlaps in the job roles of people within the AEC industries and a lot of projects combine the skills and input of all of them.

haute couture 高级时装

cloud-based CAD software 基于云计算的计算机辅助设计

Iris van Herpen 艾里斯·范·荷本（著名时装设计师）

Crystallization Iris van Herpen 与伦敦建筑师 Daniel Widrig 跨界合作的第一个 3D 打印系列作品

New words and phrases

primarily /praɪˈmerəli/   *ad.* for the most part; mainly 主要地；根本地

application /ˌæplɪˈkeɪʃn/   *n.* ① a program designed to do a particular job; a piece of software 应用程序；应用软件 ② a formal (often written) request for sth, such as a job, permission to do sth or a place at a college or university 申请；请求；申请书；申请表 ③ ~ **(of sth) (to sth)** the practical use of sth, especially a theory, discovery, etc. （尤指理论、发现等的）应用，运用

encompass /ɪnˈkʌmpəs/   *v.* ① to include a large number or range of things 包含；包括；涉及（大量事物）② to surround or cover sth completely 包围；围绕；围住

keenly /ˈkiːnli/   *ad.* in a keen and discriminating manner 敏锐地

vital /ˈvaɪtl/   *a.* ① necessary or essential in order for sth to succeed or exist 必不可少的；对……极重要的 ② [only before noun] connected with or necessary for staying alive 生命的；维持生命所必需的

advent /ˈædvent/   *n.* [sing.] **the ~ of sth/sb** the coming of an important event, person, invention, etc. （重要事件、人物、发明等的）出现，到来

inroad /ˈɪnrəʊd/   *n.* ~ **(into sth)** something that is achieved, especially by reducing the power or success of sth else（尤指通过消耗或削弱其他事物取得的）进展

sketch /sketʃ/   *n.* ① a simple picture that is drawn quickly and does not have many details 素描；速写；草图 ② a short funny scene on television, in the theatre, etc. 幽默短剧；小品

v. to make a quick drawing of sb/sth 画素描；画速写

prototype /ˈprəʊtətaɪp/   *n.* the first design of sth from which other forms are copied or developed 原型；雏形；最初形态

initial /ɪˈnɪʃl/   *a.* [only before noun] happening at the beginning; first 最初的；开始的；第一的

n. ① [C] the first letter of a person's first name （名字的）首字母 ② [pl.] the first letters of all of a person's names （全名的）首字母

v. to mark or sign sth with your initials 用姓名的首字母作标记（或签名）于……

visualize /ˈvɪʒuəlaɪz/   *v.* to form a picture of sb/sth in your mind 使形象化；想象；构思；设想

virtual /ˈvɜːtʃuəl/   *a.* ① made to appear to exist by the use of computer software, for example on the Internet （通过计算机软件，如在互联网上）模拟的；虚拟的 ② almost or very nearly the thing described, so that any slight difference is not important 很接近的；几乎……的；事实上的；实际上的；实质上的

mass /mæs/   *a.* [only before noun] affecting or involving a large number of people or things 大批的；数量极多的；广泛的

n. ① [C, usually sing.] a large amount or quantity of sth 大量；许多 ② a large amount of a substance that does not have a definite shape or form 团；块；堆

impact /ˈɪmpækt/

v. to come together in large numbers; to gather people or things together in large numbers 集结；聚集

n. ① ~ **(of sth) (on sb/sth)** the powerful effect that sth has on sb/sth 巨大影响；强大作用 ② the act of one object hitting another; the force with which this happens 撞击；冲撞；冲击力

v. ① (*business* 商) ~ **(on/upon) sth** to have an effect on sth （对某事物）有影响；有作用 ② ~ **(on/upon/ with) sth** (*formal*) to hit sth with great force 冲击；撞击

secure /sɪˈkjʊə(r)/

v. ① ~ **sth (for sb/sth)** | ~ **sb sth** (*formal*) to obtain or achieve sth, especially when this means using a lot of effort （尤指经过努力）获得，取得，实现 ② to attach or fasten sth firmly 拴牢；扣紧；关严 ③ to protect sth so that it is safe and difficult to attack or damage 保护；保卫；使安全 ④ to legally agree to give sb property or goods that are worth the same amount as the money that you have borrowed from them, if you are unable to pay the money back 抵押

a. ① likely to continue or be successful for a long time 可靠的；牢靠的；稳固的 ② feeling happy and confident about yourself or a particular situation 安心的；有把握的

embrace /ɪmˈbreɪs/

v. ① (*formal*) to accept an idea, a proposal, a set of beliefs, etc., especially when it is done with enthusiasm 欣然接受，乐意采纳（思想、建议等）；信奉（宗教、信仰等） ② (*formal*) to put your arms around sb as a sign of love or friendship 抱；拥抱 ③ (*formal*) to include sth 包括；包含

precisely /prɪˈsaɪsli/	*ad.* ① exactly 准确地；恰好地 ② accurately; carefully 精确地；细心地；仔细地 ③ used to emphasize that sth is very true or obvious（强调真实或明显）正是；确实 ④ used to emphasize that you agree with a statement, especially because you think it is obvious or is similar to what you have just said（加强同意的语气）对，的确如此，一点也不错
varying /ˈveəriŋ/	*a.* marked by diversity or difference 不同的；易变的
variant /ˈveəriənt/	*n.* a thing that is a slightly different form or type of sth else 变种；变体；变形
entire /ɪnˈtaɪə(r)/	*a.* [only before noun]（used when you are emphasizing that the whole of sth is involved 用以强调）including everything, everyone or every part 全部的；整个的；完全的
shorten /ˈʃɔːtn/	*v.* make shorter than originally intended; reduce or retrench in length or duration（使）变短；缩短
seasonal /ˈsiːzənl/	*a.* ① happening or needed during a particular season; varying with the seasons 季节性的；随季节变化的 ② typical of or suitable for the time of year, especially Christmas 节令性的；适应节日需要的；（尤指）圣诞节的
modify /ˈmɒdɪfaɪ/	*v.* ① to change sth slightly, especially in order to make it more suitable for a particular purpose 调整；稍加修改；使更适合 ② to make sth less extreme 缓和；使温和
edge /edʒ/	*n.* ① a slight advantage over sb/sth（微弱的）优势 ② the outside limit of an object, a surface or an area; the part furthest from the centre 边；边缘；边线；边沿 ③ the sharp part of a blade, knife or sword that is used for cutting 刀口；刀刃；利刃
	v. ① [+ *adv./prep.*] to move or to move sth slowly and carefully in a particular direction（使）徐徐移动，渐渐移动 ② [usually passive] ~ **sth (with/in sth)** to put sth around the edge of sth 给……加边

tailor /ˈteɪlə(r)/	*v.* ~ **sth to/for sb/sth** to make or adapt sth for a particular purpose, a particular person, etc. 专门制作；订做
	n. a person whose job is to make men's clothes, especially sb who makes suits, etc. for individual customers（尤指为顾客个别定制男装的）裁缝
desire /dɪˈzaɪə(r)/	*n.* ① a person or thing that is wished for 渴望的人；渴望的事物 ② a strong wish to have or do sth 愿望；欲望；渴望
	v. (*formal*) to want sth; to wish for sth 渴望；期望；想望
collaborate /kəˈlæbəreɪt/	*v.* ① ~ **(with sb) (on sth)** ~ **(with sb) (in sth/in doing sth)** to work together with sb in order to produce or achieve sth 合作；协作 ② ~ **(with sb)** (disapproving) to help the enemy who has taken control of your country during a war 通敌；勾结敌人
innovative /ˈɪnəveɪtɪv/	*a.* introducing or using new ideas, ways of doing sth, etc. 引进新思想的；采用新方法的；革新的；创新的
customizable /ˈkʌstəmaɪzəbl/	*v.* making or changing sth to suit the needs of the owner 可订制的，可订做的
visualize /ˈvɪʒuəlaɪz/	*v.* ~ **sth (as sth)** to form a picture of sb/sth in your mind 使形象化；想象；构思；设想
notable /ˈnəʊtəbl/	*a.* ~ **(for sth)** deserving to be noticed or to receive attention; important 值得注意的；显著的；重要的
strike /straɪk/	*v.* ① to give sb a particular impression 给（某人以……）印象；让（某人）觉得 ② to come into sb's mind suddenly 突然想到；一下子想起 ③ (*formal*) to hit sb/sth hard or with force 撞；碰；撞击；碰撞
unsettling /ʌnˈsetlɪŋ/	*a.* making you feel upset, nervous or worried 令人不安（或紧张、担忧）的

precise /prɪˈsaɪs/	a. ① clear and accurate 准确的；确切的；精确的；明确的 ② [only before noun] used to emphasize that sth happens at a particular time or in a particular way（强调时间或方式等）就，恰好 ③ taking care to be exact and accurate, especially about small details 细致的；精细的；认真的；一丝不苟的
intricate /ˈɪntrɪkət/	a. having a lot of different parts and small details that fit together 错综复杂的
innovation /ˌɪnəˈveɪʃn/	n. ① [U] the introduction of new things, ideas or ways of doing sth（新事物、思想或方法的）创造；创新；改革 ② [C] a new idea, way of doing sth, etc. that has been introduced or discovered 新思想；新方法
dimensional /daɪmenʃənl/	a. of or relating to dimensions 尺寸的
medallist /ˈmedəlɪst/	n. a person who has received a medal , usually for winning a competition in a sport（通常指体育比赛的）奖牌获得者
confirm /kənˈfɜːm/	v. ① to state or show that sth is definitely true or correct, especially by providing evidence （尤指提供证据来）证实，证明，确认 ② ~ sth~ sb (in sth) to make sb feel or believe sth even more strongly 使感觉更强烈；使确信
comm /kɒm/	abbr. (short for communication) 通信
burgeon /ˈbɜːdʒən/	v. (formal) to begin to grow or develop rapidly 激增；迅速发展
spread from ... to ...	become distributed or widespread from ... to ... 从……到……的传播
have impact upon	to affect or influence sb/sth 有影响力
as such	with respect to its inherent nature; as well; in the same way 同样地；本质地
build upon	to base sth on sth 指望；依赖

as opposed to	being in opposition or having an opponent 与······截然相反
courtesy of	with the official permission of sb/sth and as a favour 承蒙······的允许（或好意）
aim to do	try or plan to achieve sth 目的，目标
on the horizon	upcoming, forthcoming 即将到来的

▪ Reading Comprehension

Understanding the text

Answer the following questions.

1. According to the passage, which field has felt the rise of CAD over the last few decades?

2. Why do we say that the relationship between fashion and CAD is closer than ever before?

3. In the 19th century, how did designers design?

4. What does the sentence "CAD programs now increasingly used in fashion to create sketches, prototypes and designs" mean?

5. What can designers do with CAD in the fifth paragraph?

6. Compared with the traditional design method, what can designers do with CAD?

7. With seasonal trends and consumer demands to contend with, what can give fashion designers an edge?

8. In the world of haute couture, who is the most notable fashion designer to adopt 3D technology?

9. According to the examples of 3D-printed footwear, do you think 3D technology is necessary? Why or why not?

10. Do you think 3D-printed clothing will become common in the future? Please explain your reason.

Critical thinking

I. Make a Presentation.

3D technology is prevailing nowadays. Please give one or two examples to illustrate the benefits of 3D technology and try to prepare for a presentation in class.

II. Work in pairs and discuss the following questions.

1. Do you think that CAD drawing plays an important role in the process of fashion design? How does it work?

2. Do you think there will be any software better than CAD in the future? In what way?

◼ Language Enhancement

Words in use

Fill in the blanks with the words given below. Change the form when necessary. Each word can be used only once.

primarily	secure	keenly	sketch	strike
initial	virtual	embrace	shorten	confirm

1. The designer is making a _____ for his next dress.
2. When receiving the offer, Elise's _____ reaction is to decline it.
3. Finally, he _____ himself a place in the company.
4. Starting from this month, the high-speed railway will _____ the travel time from Xi'an to Chengdu.
5. This kind of clothing is _____ functional and only secondarily decorative.
6. She has remained _____ interested in international affairs since she was a child.
7. His guilty behavior _____ my suspicions.
8. It is a serious mistake for us to _____ the plan so unthinkingly.
9. It suddenly _____ me how we could improve the situation.

10. In the _____ classroom, students can communicate with their teachers through the platform.

Banked cloze

Fill in the blanks by selecting suitable words from the word bank. You may not use any of the words more than once.

A. scan	F. garments	K. necessary
B. show	G. potential	L. adopts
C. accessories	H. create	M. textile
D. tend	I. prospect	N. customization
E. perfectly	J. tailored	O. printing

The 3D printing industry is still young, and very much in the experimental phase, especially when it comes to a field such as textiles. 1. _____ organic products such as cotton and other natural fabrics is challenging, so 3D printed 2. _____ are often made from other materials like plastic. They tend to be bulky and stiff, more suited to a fashion show than to everyday wear. For these reasons, 3D printing technology has been more focused on 3. _____ such as jewelry and eyewear. These objects do not need to have the same pliability as textiles.

One such use for this technology in the clothing industry is for insoles. Companies like Canadian start-up Wiivv Wearables（一家能用 3D 技术打印鞋垫的公司）and New York-based Feetz（一家能用 3D 技术打印鞋的公司）allow the customer to 4. _____ their feet in order to 5. _____ an accurate model, from which the company 6. _____ 3D printing technology to make an insole 7. _____ fitted to the individual. This allows for cheaper 8. _____, making the insoles more accessible to consumers. While Wiivv is currently focusing on insoles, it hopes to use this technology for clothing in the future. The idea is to make 9. _____, well-fitting clothing more affordable.

As the 3D printing industry grows, so does the 10. _____ of 3D textile design.

While we are clearly still some time away from this becoming mainstream, it has a great deal of potential in the market.

Expressions in use

Fill in the blanks with the expressions given below. Change the form when necessary. Each expression can be used only once.

spread from ... to ...	the advent of	as such	as opposed to
have impact on	courtesy of	on the horizon	aim to

1. _____ 3D technology makes it possible to accomplish this task.
2. Indeed, greater inflation _____ could prove too much for the consumer to absorb.
3. A newly study has shown that loneliness can _____ one person _____ another, like a disease.
4. You can read the two handwritten documents for yourself, _____ *The Mercury News*.
5. It would be better, if they were handed cash _____ coupon.
6. The new president said his recovery plans _____ create new jobs that cannot be outsourced.
7. They may be born to perform certain things and _____ they're duty bound to follow them.
8. So, over the course of time this could _____ a very substantial _____ the community.

■ Translation

I. Translate the following paragraph into Chinese.

Using CAD software, a fashion designer can create new sketches, patterns, prints more quickly and precisely. With the increasing use of CAD, fashion designers can create multiple variations of a single design and style and adapt it to varying material and pattern. CAD also allows the fashion designer to explore various color-ways of a print developed as opposed to the traditional design methods and the alterations and rectifications are quick as well as more accurate, as it reduces rooms for errors. As the whole process from print development to collection development can be done on virtual version, very less part of the whole process is an actual physical raw materials, so the cost of collection development in fashion designing can be reduced to a minimum.

Even though we speak highly about the use and importance of CAD in the fashion industry, one should not forget and ignore the beauty and intricacy that is the essence of in manual designing and hand work. The technologies used today will be obsolete in a few years and might be replaced with a new one, but the knowledge gained and understanding developed with the help of manual designing is invaluable and hence a fine balance between the use of computer technology and manual designing is very important.

II. Translate the following paragraph into English.

    3D 打印在时尚界的使用终于到了标准化的地步。3D 打印在时尚界的主要价值在于它对普通公众的可用性和可及性，从而向大规模市场开放。3D 打印使专业人士能够超越设计的任何界限，因为它让他们有机会将最不可能的项目变成现实。个人电脑的出现使现代设计师将图案和风格的创造力提升到了一个新的水平。与使用传统工具相比，个人电脑带来了巨大的进步。单是 CAD 软件的采用就让设计师们可以画出和想象出新的款式。最近，添加制造引发了生物工程材料使用的一场革命，比如由微生物制成的皮革。如果运行 CAD 软件的台式机与我们过去 50 年所用的工具相比看起来很有新意的话，那就拭目以待未来吧。

Paragraph Writing

How to write an argumentative essay

In this unit you will learn how to write an argumentative essay, an attempt to convince others that they should agree with an idea or take a specific action.

You use arguments in your daily life more than you realize such as writing an application to a school or for a job offer. An argumentative essay presents reasons and arguments to convince people and has the same basic parts as other essays. However, its most important feature may be new to you: the presentation of convincing reasons and explanations.

An argumentative essay must provide specific and convincing evidence that supports the thesis statement. Often it is necessary to go beyond your knowledge and experience. You may need to research your topic or interview people who are experts on your topic or directly involved with it. After choosing your position and generating ideas, you need to make three major decisions: What will you include in the thesis statement? What evidence will you include? In what order will you present the evidence?

Your thesis statement should identify the issue, state your position on it, and, if possible, state the supporting points to help the readers know what to expect. Also, the thesis statement often uses the modal verbs *must, would*, and *should*. For example, this thesis statement makes it clear that the writer will argue against the use of animals for medical research: The use of animals in medical research should be banned because it is cruel, unnecessary, and disrespectful of animals.

When you write an argumentative essay, your goal is to provide the readers with evidence that they cannot easily refute, disagree with or prove wrong. So it is important to evaluate each piece of evidence you include. You may ask such a question: How might a reader reject this information? The most commonly used types of evidence are facts and statistics, quotes from authorities, and specific examples.

In addition, the argumentative essay will sound weak and one-sided if you don't respond to opposing viewpoints. Doing this, you can first concede or agree with what is seemingly right. Then you refute it using strong evidence. Responding to these opposing viewpoints in a positive manner will give you the chance to explain why your point of view is more convincing. Similarly, try to build and give the strongest conclusion, where you summarize your view and include a call to action, thereby drawing the readers into seeing your viewpoint more clearly and impressively.

Now, let's look at an example and see how to write an argumentative essay.

Animal testing is largely used in medical experiments today. Unfortunately, many animals are seriously harmed in the process. Even though animals have made a large contribution to modern medicine, animal testing should be stopped because it is immoral, unreliable, and unnecessary.

First of all, animal testing is usually cruel and immoral. On experiments, innocent animals are going through so much pain and horror. It is absolutely wrong because animals should be free to live as any other creatures and they shouldn't be taken away from their homes and natural habitats. Many animal tests are performed without any painkillers. Consequently, they lead to animals' constant pain.

In addition to being cruel, animal tests aren't reliable. People and animals are different, and that's why a medicine that appears safe for animals may not be safe for people or vice versa. One of the examples is aspirin, which is poisonous to rats and mice but not to people in some cases. While many say that animal testing allows medicines to be created that will aid human life, there are many examples showing they also cause horrible deaths in humans.

It is true that some drugs can be used on both animals and humans, but this does not mean that they have to be tested on animals in the first place when alternative methods are available. They include, for example, stem cell research. Living tissues can be grown in test tubes and new drugs can be tested on these. Computers can also be programmed to show how medicines will react in the human body.

In conclusion, inhumane, non-effective animal testing should be stopped immediately when tests could be done with alternative methods. Animals have the same rights as human beings. We can help save animals from suffering and death by donating only to medical research which doesn't experiment on animals.

Topic		Ban on animal testing
Introduction	Thesis statement	Even though animals have made a large contribution to modern medicine, animal testing should be stopped because it is immoral, unreliable, and unnecessary.
Body	Reason 1	Immoral
	Reason 2	Unreliable
	Reason 3	Unnecessary
Conclusion		Inhumane, non-effective animal testing should be stopped immediately when tests could be done with alternative methods.

Paragraph 1 is the introduction of the essay, serving as thesis statement animal testing should be stopped. Then from paragraph 2 to paragraph 4, the author provides three reasons why animal testing should be stopped. In last paragraph, the author makes a conclusion, summarizes the viewpoint and gives a call to action.

When we write an argumentative essay, our main purpose is to convince the readers that our particular view or opinion on a controversial issue is correct. At times, we may have a second purpose for our argumentative essay: to persuade our readers to take some sort of action. In a general way and as with all essays, we do the same thing: We make a point and then support it. The difference here is that argumentation advances a controversial point, a point that at least some of the readers will not be inclined to accept.

To convince readers in an argumentative essay, it is important to provide them with a clear main point and plenty of logical evidence to back it up. Because argumentation assumes controversy, we have to work especially hard to convince the readers of the validity of our position. The following five strategies may help us win over readers whose viewpoints may differ from ours.

1. Use tactful, courteous language.
2. Point out common ground.

3. Acknowledge differing viewpoints.

4. When appropriate, grant the merits of differing viewpoints.

5. Rebut differing viewpoints.

Structured writing

Read the sample essay and see how the argumentation is developed.

Sample essay

Topic

Why should you eat breakfast?

Introduction

Thesis statement: Eating breakfast is very important, especially for students.

Body

Argument 1: You should eat breakfast for your health before going to school.

Argument 2: You need food to be in a better mood.

Argument 3: You need food to do well in your classes.

Conclusion

Breakfast is the most important meal of the day.

A lot of people, especially young people, go through the day without having breakfast. Many people believe that it is not necessary, or they say that they don't have time for that, and begin their day with no meal at all. I believe that everyone should eat breakfast before going to their activities. Having breakfast is extremely important, especially for students who need energy to survive heavy schoolwork.

The first reason why you should eat breakfast before going to school is for your health. When you skip breakfast and go to school, you are looking for a disease because it's not healthy to have an empty stomach. Breakfast provides you with the energy source you need to face the day. If you don't eat breakfast, your body is weakened and you are likely to get sick, and any type of disease will have a stronger effect on you than on people who have breakfast every day.

Another reason for eating breakfast is that it can put you in a better mood. Studies show that people who eat breakfast tend to be in good moods, and those who don't eat breakfast are more easily to be tired and irritable, so eating breakfast can make you happier as it can improve your mood and reduce stress levels. Breakfast gets you started on the right track for the day.

The last reason to have breakfast every day is that you need food to do well in your classes. Your body and your brain are not going to function as well as they could because you have no energy and no strength. When you try to learn something with nothing in your stomach, you are going to have a lot of trouble succeeding. Many people think that they should not eat because they are going to feel tired, but tha's not true. Breakfast is not a very big meal, and on the contrary, you're going to feel tired if you don't have breakfast because you have spent the entire previous night without food.

You have to realize that breakfast is the most important meal of the day, and you cannot skip it without consequences for your health, your mood and your academic performance. It is better to wake up earlier and have a good breakfast than to run to school without eating anything. It is time for you to do something, and eating breakfast is the best way to start your day.

Write an essay of no less than 150 words on the topic "Why a pet?". You can follow the outline given below.

Topic:

Why a pet?

Introduction:

Thesis statement: There are some good reasons why you should keep a pet to have a positive life.

Body:

Reason 1: Give you unconditional love and companionship.

Reason 2: Increase your physical activity and improve your health.

Reason 3: Teach you to have a sense of responsibility.

Conclusion:

Everyone should keep a pet because pets provide you with unconditional love, help you retain good health, and teach you to be responsible.

More topics:

Topic 1: Drunk drivers should be imprisoned on their first offense.

Topic 2: Should girls ask boys out?

Text B

How Technology Is Changing the Fashion Industry

1   Technology has been changing our society in many ways. In recent

years, we've seen how technology has **dabbled** in the world of fashion. The Internet, **e-commerce**, **virtual** reality, **wearables**, and many other technologies are shaping this industry in **a myriad of** ways.

2 The **digital** era has **paved the way for** the **evolution** of the fashion industry. From supply chain management to marketing and sales, through design and production, the impact of technology is **undeniable**. Businesses **are forced to** adapt to these technological developments or be **rendered obsolete** and uncompetitive. Here we **take a look at** how some technologies are changing the fashion industry.

3 Social media has become a part of our globally connected world in **multiple** ways, and most recently the fashion industry is seeing a major shift in its **inspiration** for designs and trends, all thanks to social media and blogging sites. Facebook, Instagram, Twitter, Snapchat and many other social channels have a tremendous power in influencing consumer behavior and decision-making process. A study noted that 35 percent of **millennial** women **consider** social media **as** the top influencer **when it comes to** clothing purchases.

4 The way fashion models are chosen has also changed. For example, Kendal Jenner has been tapped by an international cosmetic company as their face not entirely because of her talent or **allure** but due to her millions of followers on Instagram and Twitter. The **immense** influence of social media has compelled fashion businesses to jump **on the bandwagon**.

5 Over recent years, social networking channels, particularly Facebook, have also created **personalized** data-driven advertising that specifically targets ads to consumers **most likely** to make a purchase. This has enabled fashion **labels** to reach a wider audience.

6 3D printing hasn't changed the fashion industry yet, but we may well be **on the cusp of** a revolution, something that will turn the entire industry on its head. If 3D printing prevails and becomes economically sustainable for the average home, we may see something **akin to** what happened when

the MP3 arrived in the music industry. Just as people started sharing music online, they can just as easily share designs online, print them off, and even sell them—all without having to pay a penny. There will be benefits to 3D printing; the **carbon** footprint will drop **dramatically** as shipping decreases, the manufacturing process will become faster and use less materials, and designers can test and spread their ideas across the globe **in minutes**. Only time will tell what will happen. 3D printing may never prevail, or printing your own clothes might be socially embarrassing. Perhaps, like the music industry, it may create a completely new business model in order to deal with the changes. No matter what happens, it's going to be fun.

7   Online shopping is the new **norm** and fashion retailers need to adapt to it. Establishing an online presence, setting up an e-commerce site, launching an APP, digital marketing, and other e-commerce activities have become **indispensable**. Alongside the migration of **transactions** to the digital space, fashion brands and retailers need to ensure an efficient IT **infrastructure**. An IT specialist at Bedrock IT company from Ottawa **emphasized** that having a reliable IT system is critical in ensuring seamless operations of any business, including those in the fashion industry. Through e-commerce, brands can reach consumers on a global scale. There are a lot of low-cost ways for new and emerging brands to tap their target audience. For the consumers, e-commerce has made retail purchases easier, **hassle**-free and more convenient. It has also made purchasing of international labels more accessible wherever you are in the world.

8   In less than a decade, smartphones have become a game-changer. Mobile **gadgets** have made the Internet widely available to users. **Majority of** consumers now have smartphones. They can shop, purchase and review items just with the use of their smartphones. To be able to reach this growing consumer base, it is **crucial** for fashion retailers to establish their presence in the mobile world.

9   Speaking of virtual shopping, a **drawback** with online fashion retail is the fact that consumers cannot physically fit the clothing. A variety of companies are now developing technologies that would enable an ever-more **precise** way to measure the buyer's **dimensions**. Some companies are even working on creating digital stores that allow you to tour their retail shop in 3D.

10   While the Project Glass is **tepidly** received in the market, other gadgets in the wearables category are gaining more followers and users. As such, wearable technology is seen to **eclipse** with the fashion industry. These wearable gadgets hope to free us from the endless **swiping** and checking of our smartphones. From a bluetooth-enabled cocktail ring to smartwatches to tech-embedded jewelry, we can expect the fashion world to develop innovative new devices that will make our digital lives more seamless.

11   Overall, technology has completely revolutionized the fashion industry. From how clothing is designed through how it is produced to how we buy it, the impact of technology is hard to be ignored. As new technologies emerge, consumers can expect a more personalized experience that will **meet their needs**.

12   Technology is transforming the fashion world in ways we never imagined before. And as technology continues to **evolve**, we can only expect **bolder** and better evolution in the future. For fashion retailers, it's high time to **embrace** new technologies to stay competitive.

Notes

Snapchat is an American multimedia messaging app developed by Snap Inc., originally Snapchat Inc. One of the principal features of Snapchat is that pictures and messages are usually only available for a short time before they become inaccessible to their recipients.

Bedrock IT company provides all-in IT support to small, midsized and enterprise-level companies in Ottawa. The goal of the company is to aid businesses leverage technology while lowering operational costs and increasing productivity. The time-tested methodologies and up-to-date techniques help the clients improve their overall performance and use technologies to its fullest capacity.

Google Glass, or simply Glass is an optical head-mounted display designed in the shape of a pair of glasses. It was developed by X with the mission of producing a ubiquitous computer. Google Glass displays information in a smartphone-like, hands-free format. Wearers communicate with the Internet via natural language voice commands.

Kendal Jenner　肯达尔·詹娜（模特）

New words and phrases

dabble /ˈdæbl/	*v.* ① ~ **(in/with sth)** to take part in a sport, an activity, etc. but not very seriously 涉猎；涉足；浅尝　② ~ **sth (in sth)** to move your hands, feet, etc. around in water 玩水；嬉水
e-commerce /iːkɒmərs/	*n.* commercial transactions conducted electronically on the Internet 电子商务
virtual /ˈvɜːtʃuəl/	*a.* ① made to appear to exist by the use of computer software, for example on the Internet （通过计算机软件，如在互联网上）模拟的，虚拟的　② almost or very nearly the thing described, so that any slight difference is not important 很接近的；几乎……的；事实上的；实际上的；实质上的
wearable /ˈweərəbl/	*n.* (usually used in plural) a covering designed to be worn on a person's body 衣服 *a.* (of clothes, etc. 衣服等) pleasant and comfortable to wear; suitable to be worn 穿戴舒适的；可穿戴的；适于穿戴的

digital /'dɪdʒɪtl/   *a.* ① using a system of receiving and sending information as a series of the numbers one and zero, showing that an electronic signal is there or is not there 数字信息系统的；数码的；数字式的 ② (of clocks, watches, etc. 钟表等) showing information by using figures, rather than with hands that point to numbers 数字显示的

evolution /ˌiːvəˈluːʃn/   *n.* ① (*biology* 生) the gradual development of plants, animals, etc. over many years as they adapt to changes in their environment 进化 ② the gradual development of sth 演变；发展；渐进

undeniable /ˌʌndɪˈnaɪəbl/   *a.* true or certain; that cannot be denied 不可否认的；确凿的

render /'rendə(r)/   *v.* ① (*formal*) to cause sb/sth to be in a particular state or condition 使成为；使变得；使处于某状态 ② ~ **sth (to sb/sth)** | ~ **(sb) sth** (*formal*) to give sb sth, especially in return for sth or because it is expected 给予；提供；回报 ③ (*formal*) to present sth, especially when it is done officially 递交；呈献；提交 ④ (*formal*) to express or perform sth 表达；表演；演示 ⑤ ~ **sth (as sth)** | ~ **sth (into sth)** to express sth in a different language（用不同的语言）表达；翻译；把……译成

obsolete /'ɒbsəliːt/   *a.* no longer used because sth new has been invented; out of date 淘汰的；废弃的；过时的

multiple /'mʌltɪpl/   *a.* [only before noun] many in number; involving many different people or things 数量多的；多种多样的

n. (*mathematics* 数) a quantity that contains another quantity an exact number of times 倍数

inspiration /ˌɪnspəˈreɪʃn/	*n.* ① ~ **(for sth)** a person or thing that is the reason why sb creates or does sth 启发灵感的人（或事物）；使人产生动机的人（或事物）② [U] ~ **(to do sth)** | ~ **(for sth)** the process that takes place when sb sees or hears sth that causes them to have exciting new ideas or makes them want to create sth, especially in art, music or literature 灵感 ③ [C, usually sing.] ~ **(to/for sb)** a person or thing that makes you want to be better, more successful, etc. 鼓舞人心的人（或事物）
millennial /mɪˈleniəl/	*a.* of or relating to a millenniam 一千年的；千禧年的
allure /əˈlʊə(r)/	*n.* [U] (*formal*) the quality of being attractive and exciting 诱惑力；引诱力；吸引力
immense /ɪˈmens/	*a.* extremely large or great 极大的；巨大的
personalized /ˈpɜːsənəlaɪzd/	*a.* made for or directed or adjusted to a particular individual 个性化的；个人化的
label /ˈleɪbl/	*n.* ① a piece of paper, etc. that is attached to sth and that gives information about it 标签；签条；标记 ② (*disapproving*) a word or phrase that is used to describe sb/sth in a way that seems too general, unfair or not correct （不恰当的）称谓；绰号；叫法 *v.* ① to fix a label on sth or write information on sth 贴标签于；用标签标明 ② ~ **sb/sth (as) sth** to describe sb/sth in a particular way, especially unfairly （尤指不公正地）把……称为
carbon /ˈkɑːbən/	*n.* ① [U] (*symbol* C) a chemical element. 碳 ② [C] a piece of carbon paper 复写纸
dramatically /drəˈmætɪkli/	*adv.* remarkably, markedly 剧烈地；明显地；引人注目地

norm /nɔːm/   *n.* a situation or a pattern of behaviour that is usual or expected 常态；正常行为

indispensable /ˌɪndɪˈspensəbl/ *a.* ~ **(to sb/sth) | ~ (for sth/for doing sth)** too important to be without 不可或缺的；必不可少的

transactions /trænˈzækʃn/   *n.* ① [C] ~ **(between A and B)** a piece of business that is done between people, especially an act of buying or selling （一笔）交易；业务；买卖 ② [U] ~ **of sth** (*formal*) the process of doing sth 办理；处理

infrastructure /ˈɪnfrəstrʌktʃə(r)/   *n.* [C, U] the basic systems and services that are necessary for a country or an organization to run smoothly, for example buildings, transport and water and power supplies （国家或机构的）基础设施；基础建设

emphasize /ˈemfəsaɪz/   *v.* ① to give special importance to sth 强调；重视；着重 ② to make sth more noticeable 使突出；使明显 ③ to give extra force to a word or phrase when you are speaking, especially to show that it is important 重读，强调（词或短语）；加强……的语气

hassle /ˈhæsl/   *n.* ① a situation that is annoying because it involves doing sth difficult or complicated that needs a lot of effort 困难；麻烦 ② a situation in which people disagree, argue or annoy you 分歧；争论；烦恼

gadget /ˈgæˌdʒɪt/   *n.* a device or control that is very useful for a particular job 小配件；小工具

crucial /ˈkruːʃl/   *a.* ~ **(to/for sth) | ~ (that ...)** extremely important, because it will affect other things 至关重要的；关键性的

drawback /ˈdrɔːbæk/   *n.* a disadvantage or problem that makes sth a less attractive idea 缺点；不利条件

precise /prɪˈsaɪs/   *a.* ① clear and accurate 准确的；确切的；精确的；明确的 ② [only before noun] used to emphasize that sth happens at a particular time or in a particular way（强调时间或方式等）就，恰好 ③ taking care to be exact and accurate, especially about small details 细致的；精细的；认真的；一丝不苟的

dimension /daɪˈmenʃn/   *n.* ① [usually pl.] the size and extent of a situation 规模；程度；范围 ② a measurement in space, for example the height, width or length of sth 维（构成空间的因素）；尺寸 ③ an aspect, or way of looking at or thinking about sth 方面；侧面

tepidly /ˈtepɪdli/   *ad.* in an unenthusiastically lukewarm manner 不冷不热地，微温地

eclipse /ɪˈklɪps/   *v.* ① to make sb/sth seem dull or unimportant by comparison 使失色；使相形见绌；使丧失重要性 ① [often passive] (of the moon, the earth, etc. 月球、地球等) to cause an *eclipse* 遮住……的光 ① [sing.U] a loss of importance, power, etc. especially because sb/sth else has become more important, powerful, etc. （重要性、权势等的）丧失；黯然失色；暗淡 ② [C] an occasion when the moon passes between the earth and the sun so that you cannot see all or part of the sun for a time; an occasion when the earth passes between the moon and the sun so that you cannot see all or part of the moon for a time 日食；月食

swipe /swaɪp/   *v.* to pass a plastic card, such as a credit card, through a special machine that is able to read the information that is stored on it 刷（磁卡）

n. ① an act of hitting or trying to hit sb/sth by swinging your arm or sth that you are holding 抡打；挥击 ② an act of criticizing sb/sth 批评；抨击

evolve /iˈvɒlv/	*v.* ① ~ **(sth) (from sth) (into sth)** to develop gradually, especially from a simple to a more complicated form; to develop sth in this way（使）逐渐形成；逐步发展；逐渐演变 ② (*biology* 生) ~ **(from sth)** (of plants, animals, etc. 动植物等) to develop over time, often many generations, into forms that are better adapted to survive changes in their environment 进化；进化形成
bold /bəʊld/	*a.* ① (of people or behavior 人或举止) brave and confident; not afraid to say what you feel or to take risks 大胆自信的；敢于表白情感的；敢于冒险的 ② (of shape, colour, lines, etc. 形状、颜色、线条等) that can be easily seen; having a strong clear appearance 明显的；轮廓突出的 ③ (*technical* 术语) (of printed words or letters 印刷字或字符) in a thick, dark type 粗体的；黑体的
	n. [U] (*technical* 术语) thick, dark type used for printing words or letters 黑体；粗体
embrace /ɪmˈbreɪs/	*v.* ① (*formal*) to accept an idea, a proposal, a set of beliefs, etc., especially when it is done with enthusiasm 欣然接受；乐意采纳（思想、建议等）；信奉（宗教、信仰等） ② (*formal*) to put your arms around sb as a sign of love or friendship 抱；拥抱 ③ (*formal*) to include sth 包括；包含
a myriad of	many, a lot of 大量的
pave the way for	smooth the way for; creat the circumstances to enable (something) to happen or be done 为……铺平道路；为……做好准备
be forced to	be pushed to; have to 被迫；不得不
take a look at	examine，inspect 检查，看一眼
consider ... as	think of sb/sth in a particular way 认为；以为；觉得

when it comes to	refer ... to 当提到……；就……而论
on the bandwagon	在有胜利希望的一边
most likely	possibly, likely 很有可能
on the cusp of	at the point in time that marks the beginning of sth 在（某种事物）开始的时间点上
akin to	similar to or related to in quality or character 类似的
in minutes	in a few minutes 在几分钟内
majority of	most of 大多数，大部分
meet one's needs	satisfy what is needed or what sb asks for 满足某人的需要；符合某人的需要

▬ Reading Comprehension

Understanding the text

Answer the following questions.

1. What has paved the way for the evolution of the fashion industry?

   A. The technologies has paved the way for the evolution of the fashion industry.

   B. The internet has paved the way for the evolution of the fashion industry.

   C. The wearables has paved the way for the evolution of the fashion industry.

   D. The digital era has paved the way for the evolution of the fashion industry.

2. The fashion industry is seeing a major shift in its inspiration for designs and trends, all thanks to _____.

   A. social media

   B. blogging sites

   C. social media and blogging sites

   D. None of them

3. Kendal Jenner has been tapped by an international cosmetic company as their face because of _____.

   A. her talent

B. her allure

C. her millions of followers on Instagram and Twitter

D. her appearance

4. According to the paragraph 6, what do you know about 3D printing?

   A. 3D printing has changed the fashion industry.

   B. 3D printing has become economically sustainable for the average home.

   C. By using 3D printing, designers can test and spread their ideas across the globe in minutes.

   D. 3D printing may never prevail, it may create a completely new business model in order to deal with the changes.

5. Through _____, brands can reach consumers on a global scale.

   A. online shopping

   B. online presence

   C. fashion retailer

   D. e-commerce

6. What do you know about virtual shopping?

   A. A shortcoming is that consumers cannot physically fit the clothing.

   B. A few firms are willing to develop technologies that would enable a precise way to measure the buyer's dimensions.

   C. A variety of companies are working on creating digital stores that allow consumers to visit their retail shop in 3D.

   D. For its convenience, consumers are willing to do virtual shopping.

7. While the Project Glass _____ in the market, other gadgets in the wearables category are gaining more followers and users.

   A. is popular

   B. is not popular

   C. has many followers

   D. has many users

8. What technologies have changed the world of fashion?

   A. Internet.

   B. E-commerce.

   C. Virtual reality.

   D. All of the above

Critical thinking

Work in pairs and discuss the following questions.

1. How do you understand the sentence "The digital era has paved the way for the evolution of the fashion industry"?
2. Apart from the technologies mentioned in this passage, are there any other technologies can influence the world of fashion? What are they?

◤ Language Enhancement

Words in use

Fill in the blanks with the words given below. Change the form when necessary. Each word can be used only once.

dramatically	crucial	indispensable	emphasize	bold
norm	dimension	gadget	evolve	drawback

1. Smartphones have become an _____ part of our daily life.
2. Her new position added new _____ to her job.
3. The quality of people's life has improved _____ since 1949.
4. Families of six to ten are the _____ in 1950s.
5. This story _____ that one should not boast his strengths to others.
6. It was a _____ move on their part to open a business in France.
7. How did apes _____ into humans?
8. We need a simple _____ to finish this project.
9. The major _____ of this new system can not be solved.
10. Your approval is _____ to the accomplishment of our objectives.

Expressions in use

Fill in the blanks with the expressions given below. Change the form when necessary. Each expression can be used only once.

pave the way for	be forced to	take a look at	when it comes to
most likely	on the cusp of	in minutes	be crucial for

1. The _____ reason for failure is the deficit of language.

2. _____ to drawing sketches, no one can compare with him.

3. Mastering good language skills _____ communication.

4. The technology can scale easily as patients can do the tests themselves _____ and are as cheap as making a local phone call.

5. Thanks to everyone's effort, our team is _____ victory.

6. Let's _____ a case study before moving on to the new unit.

7. Obviously, the discussions are meant to _____ for further negotiations.

8. The company will _____ take responsibility for the loss of its employees.

Sentence structure

I. Complete the following sentences by translating the Chinese into English, using "When it comes to ..." structure.

Model:	A study noted that 35 percent of millennial women consider social media as the top influencer _____ (当谈到买衣服的时候).
	→A study noted that 35 percent of millennial women consider social media as the top influencer <u>when it comes to clothing purchases</u>.

1. _____(当谈论到污染时), industrial pollution is a major offender.

2. _____(当谈论到 影星的隐私时), people cannot help gossiping, especially the young.

3. _____(当谈论到

学习英语时), most people hold that one should focus on learning the ability of listening and speaking.

II. Rewrite the following sentences by using "Not only ..., but also ...".

Model: This not only saves time and reduces errors, but it also cuts costs down by letting designers view virtual versions of finished products without needing to purchase materials.

→Not only does this save time and reduce errors, but it also cuts costs down by letting designers view virtual versions of finished products without needing to purchase materials.

1. She not only wrote lyrics for songs, but she also composed the music.

_____
_____.

2. Smartphone not only appeals to children, but it also attracts adults.

_____
_____.

3. The man had not only been fined, but he also been put in prison.

_____
_____.

Extensive Reading

Social Media Influences on Fashion

Social media has become a part of our globally connected world in multiple ways, and most recently the fashion industry is seeing a major shift in its inspiration for designs and trends, all thanks to social media and blogging sites.

In the fashion industry, social media platforms can be used to interact with the consumer, as a means of networking with others in the industry, and as a way of building an online presence. But social media can also influence designers in unique ways that are changing the way many designers create their fashions.

Designers and major players in the fashion industry find inspiration for their designs and creations in many places: from recent vacations to visits to fashion capitals of the world, but the Internet and social media sites have created an entirely new approach that is getting both praise and criticism from the fashion community. On one side, some see the Internet and social media sites as more of a secondary resource for inspiration, with the fear that these resources could "dumb down" the inspirational process and take away from its depth. On the other side, designers know that their target customers are on social media sites and want to be heard and be a part of the inspiration processes themselves. Either way, social media has, and is, changing the fashion industry and it is anyone's guess what the shift will continue to look like.

One major shift that social media has had is simply that the average person can now influence fashion in ways never before possible. Just consider the recent Zac Posen's Spring-Summer Ready to Wear 2015 Collection; influenced by comments and suggestions from his over 640,000 Instagram followers. It all began when Posen posted images of sunsets from a vacation on his Instagram account and followers began asking for prints in these hues, and "through the comments and pictures we got a new perspective about our creations," said Posen in a recent article in *The New York Times*. As a result, the final product includes a maxi dress in the hues of the sunset from that Instagram image as part of his 2015 collection.

It's not just the inspiration and influences on major fashion designers that social media is evolving; it's also the way we view fashion and the industry as a whole. For decades, a few big names dominated the fashion industry in a very top-down fashion but now the average fashion blogger can influence major designers via social media outlets. Reality television shows like Fashion Star and Project Runway have also lead to a shift in the way we view fashion designers, but social media has acted as a catapult to push these "unknowns" into situations where they can make their designs and ideas known to millions. Social media sites act as a platform for the average person, and major fashion designers know these people are out there so they can also reap benefits by reaching customers on a new level that is more intimate and interactive, rather than the highfalutin fashion runways.

The naturally interactive qualities of social media also make social media an effective tool, allowing people to be a part of the process of fashion making. A study considered data from top five social media networking sites during the 2013 New York Fashion Week and found that "more people are using social media for wardrobe advice, inspiration and the latest trends," and fashion related Tweets doubled from the same event just a year earlier. From people sharing Instagram photos and Tweets from the sidelines of top fashion runways, people from home can interact and engage in the fashion shows just like the attendees. This is great for the fashion industry which can tend to be closed off and secretive in their inspiration processes, opening it up for more people to both appreciate and enjoy.

And it's not just high fashion experiencing influences from social media. From popular viral videos shared over and over again on social media to "jokes" and social media inspired sayings on T-shirts, buying trends centered on hashtags and our online behavior is shifting too. According to 2014 market trends, a 2014 survey of social networks and buying behavior "showed significant proportions of shoppers being influenced by (and participating in) social sites and friends during their upcoming holiday shopping," with the results showing: 30% of shoppers had made a purchase via social media in the last year (up by 12% from 2014), 49% planning to make a purchase because of a social referral, and 44% intending to discover new products via social networks. In addition, the survey looked at social participation and found that 48% of those surveyed think it is important to share product recommendations via social media networks.

So the point is, with social media sites only growing from here and buying trends also shifting to show major influences from social media sites, it's a brave new world for the fashion industry and retailers.

Glossary

accelerate	6B	anthropology	3A
accessible	6A	apparel	3B
accessory	7A	appealing	5A
accurate	7B	application	8A
activism	6A	appreciate	5A
adapt	5B	apprentice	3B
adaptability	2A	approach	5B
address	3B	approximately	6B
adopt	6A	archaeological	1A
adornment	4A	archery	4A
advancement	5B	arena	7B
advent	8A	aristocracy	3A
aerospace	1A	aristocratic	3A
afar	3A	artificial	1A
affect	7A	asset	5A
affluent	7A	attire	3B
afloat	5B	attribute	4A
agency	5B	authenticate	6B
aggregation	2A	aviation	6B
ale	7B	awareness	5B
allure	8B	bachelor	3B
analyse	7B	back-end	7B
anatomy	3A	beneficiary	7B
angle	7B	bias	7B
animation	3A	blanket	7A
anonymous	3B	bleach	4A
anonymously	3B	blended	2A
antagonistic	3A	blurry	5B

considerable	2A	cusp	8B
consistent	5A	customize	2A
constantly	5B	cylinder	4B
construction	1B	cylindrical	4B
contract	4B	dabble	8B
contractor	6B	damask	7A
contractor	7A	debut	6B
contradiction	3A	deceleration	1B
contradictory	5B	decent	5A
controversial	6A	decompose	3A
conventional	6A	defect	2A
conversion	5B	deficiency	2B
coordinate	2B	definitely	5B
copious	6B	defy	5A
corduroy	3A	degumming	2A
corporate	7A	democratize	6B
cosmic	3A	demographic	5A
costume	3A	denote	3A
couple	7B	density	2A
couture	3A	deter	5B
couturier	7A	detrimental	6A
craft	5A	dexterity	3B
craftsmanship	7B	dialectical	3A
credible	5A	dialectically	3A
criticize	6A	dictate	7A
crocheting	1A	differentiate	4A
crucial	8B	digital	8B
cuff	2B	dignified	4A
curate	7B	dimension	8B
currently	5B	disastrous	6B
curved	2A	discipline	7B
curvy	6A	discount	7A

discrete	2B	elastic	2B
disintegration	2B	eliminate	2B
disorganized	2A	emanate	4A
dispel	3B	emancipate	4B
dispersion	2B	embarrass	8B
dispersivity	2B	emblem	4A
disposal	5B	embrace	6A
disseminate	5B	embroider	4A
distinctive	4A	emerge	6A
distinguish	4A	emission	6B
distribution	7B	emphasize	8B
diverse	6A	emulate	5B
diversified	1A	enable	7A
diversify	4B	enforce	1B
documentation	3B	engage	5A
domestically	6B	enhance	3A
dominant	6A	enterprise	1B
don	6B	era	6A
double-digit	1B	ergonomics	3A
dramatic	4A	esteem	4B
dramatically	8B	ethical	6B
drape	3A	ethics	3A
drawback	8B	ethnic	4B
dual	2A	evident	5A
dye	4A	evolution	8B
dyeing	4A	evolve	8B
eclipse	8B	exacerbate	6B
ecological	6B	exaggeration	2B
ecology	3A	excavation	1A
e-commerce	8B	exceed	1B
eddy	2A	exclusive	6A
efficacy	5A	expertise	3B

impulse	6B	knit	2B
impurities	2A	knitwear	1B
inappropriate	6A	kudzu	1A
inclusive	6A	label	8B
incorporate	7B	lace	7A
incredibly	3A	landfill	6B
incubator	7B	landscape	6A
indicator	1B	lateral	2B
indispensable	8B	layout	1B
infinite	7A	lengthy	5A
influential	6A	licensing	7A
infrastructure	8B	linear	2A
ingenious	3A	literally	4A
ingot	1A	longitudinal	2A
innovation	8A	loop	2B
innovative	8A	loosened	2A
inroad	8A	machinery	1B
insensitive	6A	mainstream	6A
insulate	7B	manufacture	7A
integrate	4A	margins	7A
intelligent	3A	market	7A
intern	3B	marketplace	7A
internship	3B	mass	8A
intersperse	6B	mature	1B
interval	6B	maximize	5A
intricate	8A	medallist	8A
intrinsic	3A	mentorship	7B
investigation	6B	merge	4B
investment	6A	mesh	2B
issue	7B	meteorology	3A
item	5A	metric	5A
jacquard	2B	millennial	8B

minimum	6B	outspoken	6A
minority	6A	outweigh	1B
miscellaneous	2A	overall	1B
misconception	3B	overcome	2B
modeling	3A	oversee	3B
modest	6A	paleolithic	1A
monitor	5A	palette	3A
motivation	7B	panel	4A
mourning	3A	participant	5B
mutual	4B	passionate	7B
multidisciplinary	7B	peat	2A
multiple	8B	pedalling	1A
mutual	3A	permeability	2B
mutually	3A	personalize	5B
myth	3B	perspective	7B
neckline	2B	Phase	1B
negatively	6A	phenomenon	7B
nomadic	4B	physiology	3A
norm	8B	pillar	1B
normal	1B	pivotal	6A
noteworthy	5A	platform	5A
oblique	2B	plaza	6B
obsolete	8B	pleated	2B
offering	7B	plush	2B
offset	6B	poised	4A
opinionated	6A	polish	3B
opt	5B	polypropylene	2B
optimize	1B	popularize	7A
outfit	7A	populate	7B
outperform	5A	port	2B
outreach	5A	portray	3B
outsource	7B	pottery	1A

precise	8b	rational	2A
preference	5B	raw	2A
preliminary	2A	recommendation	5B
present	5A	re-establish	2A
presentation	5A	reflect	6A
preserve	4B	reflection	7B
prevail	4A	reflective	7B
previous	1B	regime	4A
primarily	8A	regional	1B
primary	5A	register	7B
prime	3B	reign	4A
primitive	1A	relatively	7A
prior	3A	release	6B
process	7B	release	2A
procure	3B	religion	7B
profile	3A	remarkable	1B
profit	7B	removal	2A
profound	1B	remove	2A
progressive	6A	render	8B
prominence	7A	renewal	6B
prominent	3A	replace	2A
propensity	5B	represent	7B
prospective	3B	representative	7A
prototype	8A	reside	5B
purportedly	6B	resistant	4B
pursuit	4A	resonate	5A
pyramid	7B	retailer	6A
ramie	4A	retailer	3B
random	5B	reveal	6B
random	3A	revenue	5B
rapper	7B	reverse	4B
ratio	7B	rib	2B

rigid	4A	slit	4B
rotary	2A	solidarity	7B
rotation	2A	speciality	3B
runway	6A	specialize	3B
satin	2B	specification	2A
scene	6A	spectrum	6A
scouring	2A	spin	4A
scrutinize	6A	spin	2A
seams	2B	spindle	1A
seasonal	8A	spinnability	2A
secure	8A	splicing	2A
segment	5B	sponsor	6B
segmentation	2A	spur	4A
self-actualization	7B	stability	4A
sensitive	7B	staple	2A
sericulture	1A	startup	7B
serviceability	2B	statistics	7A
shank	4B	status	7B
shortfall	6B	stiff	2B
showcase	6A	stipulate	4B
showcase	4B	strategic	1B
showy	7A	strategy	5A
signage	7B	stunning	5A
significant	6A	subpar	5B
significantly	5A	substitute	5B
silhouette	3A	subvert	6A
silo	7B	subversive	4B
simulate	3A	succeed	4B
simultaneously	7A	superficial	7B
sketch	3B	superior	1A
slacken	1B	surge	5A
slim	3B	sustainability	7B

sustainable	1B	triangle	2A
swiping	8B	trillion	1B
symbolism	7B	trimming	7A
synthesize	3A	trinkets	7A
synthetic	6B	trough	1B
systemic	1B	tubular	3A
tactic	5A	turnover	1B
tailor	8A	twist	2A
tailoring	3B	ultimately	2B
taper	4A	unadorned	4A
temperament	4B	undergo	1B
template	2B	underlinen	4A
tepidly	8B	underscored	7A
terry	2B	undoubtedly	5B
testimonial	5A	unearthed	1A
texture	3A	uneven	2B
thoughtfully	6B	unflattering	5B
toll	6B	unification	4A
topple	4B	uniform	4A
toss	6B	unionize	6B
trace	6B	universal	7B
trademark	7A	unprecedented	1A
transactions	8B	unprecedented	1B
transfer	1B	upgrade	1B
transformation	1B	usher	4A
transgender	6A	utilize	5A
transparency	7B	varying	8A
transparent	6B	vascular	1A
transversely	2A	velvet	7A
trappings	4A	vendor	3B
treacherous	5B	version	7B
tremendous	1B	vertical	1B

veteran	5B	wearable	8B
vigor	4A	whereby	7B
virtual	8B	wholesaler	3B
virtuous	6B	witness	4A
visual	3A	wrapped	2A
visualize	8A	yarn	4A
vital	8A	yarns	2A
vitality	4A	yell	5A

a broad range of	5A	be forced to	7A
a large number of	2A	be forced to	8B
a myriad of	8B	be pointed out	2B
a roster of	7A	be processed into	2A
a variety of	2A	be referred to as	4B
a variety of	2B	be responsible for	3B
according to	6B	be separated into	2B
account for	1B	be substituted for	4B
aim to	8A	be suitable for	2B
akin to	8B	be summarized into	3A
apart from	5B	be viewed as	3A
apply for	4A	break off	2B
are apt for	3B	breaking down	7B
as much as possible	2A	build upon	8A
as opposed to	8A	care about	7B
as such	8A	carry out	1B
associate with	6B	cater to	7A
at intervals	6B	coincide with	6A
at one's disposal	5B	come into being	4A
at rock bottom prices	7A	come up with	5A
at the bottom	7B	communicate with	7B
at the same time	2A	compare with	1B
attribute to	6B	contribute to	3A
bank on	7A	contribute to	4B
be acquainted with sth	3B	coordinate with	7A
be added to	3A	courtesy of	8A
be affected by	1B	cut through	5A
be assigned by	1A	date back	7A
be associated with	3A	date back to	4A
be composed of	4A	dependent on	3A
be divided into	2A	dcrive from	4B
be favorable to	1B	deter from	5B

dispose of	6B	in particular	6A
dress up	7A	in the midst of	5B
due to	3B	incorporate … into	5A
dump into	1A	indispensable to	3A
even if	7B	keep pace with	6B
evolve into	4B	lay in	7B
evolve into	1A	lay the foundation for	4B
evolve from	3A	lead to	5B
excel at	5A	lie in	5A
expose to	7A	majority of	8B
extend to	1B	make good use of	1B
feature in	6A	matched with	7B
fit in	5A	meet one's needs	8B
flock to	7A	most likely	8B
flood with	6B	on a daily basis	4A
focus on	5A	on behalf of	5A
for the sake of	5A	on the bandwagon	8B
get closer to	7B	on the basis of	2B
get in touch with	5B	on the cusp of	8B
get/be initiated at	1A	on the horizon	8A
give birth to	4B	or so	5A
give full play to	2B	originate from	1A
happen to	5B	other than	7B
have impact on	8A	participate in	7B
in a … manner	1B	pave the way for	8B
in a … sense	2A	plaster with	7A
in addition to	3B	put forward	1A
in answer to	6A	put out	5A
in defense of	7B	reach a peak	1B
in essence	5A	reach a…level	2A
in minutes	8B	refer to	6B
in parallel	2A	regardless of	4B

rely on	5A	take a look at	8B
resonate with	5A	take over	5B
responsible for	6B	tap into	6A
result in	2A	tear into	2A
revolve around	5B	tend to be	4A
roll out	6B	thanks to	7B
set up	1A	there is no point doing	3B
shift to	4A	to a certain extent	1B
show off	5B	to begin with	2B
sign up for	5B	to sum up	1B
slow down	1B	trace back to	1A
so as to	2A	translated into	7A
speaking of	5B	turn … into …	3B
specialize in	3B	turn into	7A
spread from ... to ...	8A	vary with sth	3A
stand out	5A	when it comes to	8B
strike up	6A	with the help of	2A
substitute for	5B	year-on-year	1B